TAPPING WATER MARKETS

Tapping Water Markets is about the past, present, and future of water markets. It compares water markets with political water allocation, documents the growth of water markets, and explores the ways in which water markets can be improved and implemented further. This book provides up-to-date information of where and why water shortages are occurring and where and why water markets are evolving to resolve conflicting water uses.

Though the main focus is on the United States, it includes examples from other parts of the world to show how water markets are beginning to thrive. It contains institutional detail that is accessible to people who are not economic or hydrologic experts, and comes alive with numerous examples and case studies of water markets.

The book begins with an analysis of water institutions as they have varied over time and location. It then covers a range of discrete water management topics including surface water allocation, groundwater management, environmental flows, and water quality trading. The book concludes with predictions about the future of water scarcity and the ability of water markets to shape that future more positively.

Terry L. Anderson is the Executive Director of PERC and Senior Fellow at the Hoover Institution, Stanford University. Anderson's work helped launch the idea of "free market environmentalism" with the publication of his book by that title, coauthored with Donald Leal. Anderson is the author or editor of more than thirty books and has published widely in both professional journals and the popular press. He received his bachelor's from the University of Montana and his Ph.D. in economics from the University of Washington.

Brandon Scarborough is a Research Fellow at PERC specializing in water markets for economic and environmental applications, and the associated institutional structures that may promote or inhibit trading. Other research interests include markets for ecosystem services, climate change, and the efficacy of using forests to sequester carbon as part of a national or international abatement strategy. He holds bachelor's degrees in biology and business and a master's in applied economics.

Lawrence R. Watson is a Research Fellow and the Director of Applied Programs at PERC. He specializes in contracts for environmental resources, particularly water and wildlife, and consults to environmental organizations and state agencies on market-based conservation strategies. Watson earned a law and a master's degree from Duke University and a bachelor's degree from Clemson University.

TAPPING WATER MARKETS

Terry L. Anderson, Brandon Scarborough,
and Lawrence R. Watson

RFF PRESS
RESOURCES FOR THE FUTURE

First published 2012
by Routledge
711 Third Avenue, New York, NY 10017

Simultaneously published in the UK
by Routledge
2 Park Square, Milton Park, Abingdon, Oxon OX14 4RN

Routledge is an imprint of the Taylor & Francis Group, an informa business

Library of Congress Cataloging in Publication Data
Anderson, Terry Lee, 1946–.
    Tapping water markets/Terry L. Anderson, Brandon Scarborough,
    and Lawrence R. Watson.
    p. cm.
    1. Water-supply—Economic aspects.   2. Water transfer.
    I. Scarborough, Brandon.   II. Watson, Lawrence R.   III. Title.
    HD1691.A53 2012
    333.91—dc23   2011043335

ISBN13: 978-1-61726-099-5 (hbk)
ISBN13: 978-1-61726-100-8 (pbk)
ISBN13: 978-0-203-13607-2 (ebk)

Typeset in Bembo and Stone Sans
by Florence Production Ltd, Stoodleigh, Devon

ABOUT RESOURCES FOR THE FUTURE *AND* RFF PRESS

Resources for the Future (RFF) improves environmental and natural resource policymaking worldwide through independent social science research of the highest caliber. Founded in 1952, RFF pioneered the application of economics as a tool for developing more effective policy about the use and conservation of natural resources. Its scholars continue to employ social science methods to analyze critical issues concerning pollution control, energy policy, land and water use, hazardous waste, climate change, biodiversity, and the environmental challenges of developing countries.

RFF Press supports the mission of RFF by publishing book-length works that present a broad range of approaches to the study of natural resources and the environment. Its authors and editors include RFF staff, researchers from the larger academic and policy communities, and journalists. Audiences for publications by RFF Press include all of the participants in the policymaking process—scholars, the media, advocacy groups, NGOs, professionals in business and government, and the public.

CONTENTS

ILLUSTRATIONS

Tables

Figures

PREFACE

Terry L. Anderson

Thirty years ago, when I started thinking and writing about water markets, nearly every economist and policy analyst argued that market failure permeated water allocation, and there were very few examples of water markets at work. Hence, my first book on the subject, *Water Crisis: Ending the Policy Drought* (Cato Institute, 1983), was mostly focused on what institutional changes were necessary to give water markets a chance. Needless to say, one of the biggest hurdles to water markets was prices set by politicians and bureaucrats that made water "cheaper than dirt" (see Chapter 2). Unfortunately, that remains a hurdle today.

Problems inherent in setting water prices through the political process draw our attention to the fact that it is not so much market failure that is holding markets back, but rather it is political failure. Politics stand in the way of reforming institutions that allocate water, defining and enforcing water rights, and determining the transaction costs associated with trade. In addition to subsidies, governmental agencies do not allow willing buyers to buy from willing sellers, and laws still require "use it or lose it" thus discouraging conservation and trades. Perhaps water politics explains why the commissioner of the Bureau of Reclamation, an agency that is one of the biggest hurdles to water marketing, called me a "kiddie car economist" when I presented ideas from my book at the Department of the Interior in 1983.

Interestingly, environmentalists were some of the first to recognize the potential for water markets. My friend, the late Tom Graff, was a market pioneer at the Environmental Defense Fund (EDF) and in the environmental community, generally. Writing in the *Los Angeles Times* about the proposal to build a canal to divert water from northern to southern California in 1982, he asked, "has all future water project development been choked off by the new conservationist-conservative alliance . . . ?" Tom went on to propose reforms including eliminating legal barriers to voluntary sale and purchase of water rights and pricing water to better reflect the true cost of using the resource.

Tom represented the EDF in *National Audubon v. Superior Court* (1983), a lawsuit filed by a consortium of environmental groups to stop Los Angeles from reducing water levels in Mono Lake. The environmentalists "won" the case, meaning the court ruled that Los Angeles had to reduce its diversions from the Mono Lake watershed to bring back a more natural balance to the lake. When it took more than a decade to actually see more water flowing into the lake, Tom told me that it would have been cheaper and the environment would have been improved sooner if they had simply negotiated in the marketplace to get some of the water.

Though many economists and policy analysts continue to emphasize market failure, a growing number are recognizing what Tom Graff realized—water markets offer great potential for improving water use efficiency and water quality. This is part of a more general recognition that markets can work for environmental improvement if property rights can be more clearly defined and enforced. In the case of water, the prior appropriation doctrine provides the foundation for establishing and clarifying water rights, which is why water markets are making more progress in the western United States. Armed with more evidence of political failure in water allocation and of increased use of water markets, Pamela Snyder and I wrote *Water Markets: Priming the Invisible Pump* (Cato Institute, 1997). That book strengthened the case for and documented examples of water markets.

Since the writing of *Water Markets: Priming the Invisible Pump* in 1997, water markets have made even greater progress. While more and more scholars and environmental activists have recognized the efficacy of water markets, entrepreneurs have exploited the gains from trade. Water trusts patterned after land trusts have been formed in Oregon, Washington, Montana, and Colorado to increase stream flows for aquatic life. For-profit companies have been formed to broker water deals, to improve infrastructure, and to store and deliver water.

Given the growth in water markets over the past 15 years, I realized that the *Water Markets* book was dated and therefore started thinking about revising and updating it. As with any such endeavor, it seemed an easy task to add a few new examples, update the references, and be done with it, but it did not quite work that way.

In this case, I began by asking the Cato Institute for permission to make the revision and seek another publisher. Having "cut my teeth" on water issues with Cato, I certainly didn't want to appear ungrateful, but the people there understood that these ideas have moved beyond the thinking of those of us on the more market fringe to become quite mainstream. Therefore, Cato graciously gave me the green light and encouraged me to move forward with a revision.

Because Pam Snyder had moved on with her legal profession, I recruited Brandon Scarborough and Reed Watson, research fellows at the Property and Environment Research Center (PERC), to assist me with the project. Brandon had already done pioneering work in documenting the use of markets to enhance instream flows (Chapter 7), and Reed's law degree brought valuable human capital for better understanding the legal institutions, especially in the East. Together we dove into the project.

It did not take us long to discover that we were writing a new book, not just updating an existing one. Almost everything except the history of water institutions (Chapter 3) was changing. Water markets were on the march in the United States and around the world (Chapter 10), entrepreneurs were finding new opportunities to harness water markets for environmental improvement (Chapter 8), and even the riparian institutions in the eastern United States are evolving to accommodate some water marketing (Chapter 6).

Therefore, we offer *Tapping Water Markets* as a sequel to, rather than a revision of, *Water Rights*. A reader comparing the two will find every chapter revised to include updated data and references and will find significant new material in every chapter.

Bringing this book to fruition required support on many fronts. First, I would never have undertaken this project, let alone completed it, if Brandon and Reed had not signed on. They represent the next generation of scholars interested in pushing the frontiers of free market environmentalism, and I am proud to have my name on the cover of this book with theirs. Second, I want to thank the Cato Institute for publishing *Water Rights: Priming the Invisible Pump* and for giving us permission to use ideas and materials from that book in this one.

Completing a book like this requires significant financial support. For such support, we thank the M. J. Murdock Trust, the Searle Freedom Trust, the Earhart Foundation, the Sand County Bradley Fund for the Environment, and the Dunn's Foundation for the Advancement of Right Thinking for support they provided through PERC. The project was also supported by the John and Jean DeNault Task Force on Property Rights, Freedom, and Prosperity at the Hoover Institution, Stanford University. John and Jean are willing to invest in "ideas defining a free society," the mission of the Hoover Institution. Thanks also go to John Raisian, director of the Hoover Institution, for his entrepreneurship that created the DeNault Task Force and for encouraging so much of my research.

The progression from initial drafts to a final product required many eyes, ears, and minds beyond those of the authors. Our colleagues at PERC were always available to listen to our arguments and provide feedback. We also thank the PERC staff, which worked diligently to create the finished product. Michelle Lim was particularly patient with correcting our mistakes. Outside of PERC, Don Reisman committed RFF and Earthscan to the project, and Natalja Mortensen carried the ball for Taylor and Francis. Finally we thank, in alphabetical order, the many scholars who read and commented on all or part of the manuscript: Greg Characklis, Chris Corbin, Zachary Donohew, Martin Doyle, Monique Dutkowsky, Terry Eccles, Andy Fischer, Millie Heffner, James Huffman, Thomas Iseman, Brian Kirsch, Gary Libecap, Andrew Morriss, Elizabeth Richards, Deborah Stephenson, and Kerri Strasheim. Of course, we take full responsibility for any mistakes or deficiencies that remain.

Brandon, Reed, and I are committed to the ideals of a free and responsible society and to fostering environmental quality through such a society. We believe that free market environmentalism has tremendous potential for making this

1

WHY THE CRISIS?

If you were an alien heading toward the "Blue Planet," you might be surprised by the barrage of predictions of water shortages. From global organizations to local newspapers, water crises dominate reports and headlines even in the absence of droughts. In 2008, the United Nations estimated that 450 million people (roughly 1 in 15) were experiencing water shortages and predicted that the number could expand to 1.8 billion people (UNEP 2008). The World Health Organization estimates 884 million people rely on unimproved and often unhealthy drinking water sources, while a staggering 2.5 billion have limited or no access to improved sanitation facilities and waste disposal[1] (WHO and JMP 2008). The World Bank warns that fresh water consumption is rising quickly, and the availability of water in some regions is likely to become one of the most pressing issues of the twenty-first century (World Bank 2003, 2).

Mainstream media is similarly fond of water crisis, too, especially when there is a drought or water contamination. Newspaper headlines consistently deliver the water gloom and doom message: "Northern Kenya Pupils Miss School over Drought" (*Daily Mirror* January 3, 2011), "Kampala Water Crisis Intensifies" (*Daily Monitor*, December 22, 2010), "A National Water Crisis is on the Verge of Gushing" (*U.S. News and World Report* May 27, 2007), and "Widespread Pollution Taints Iowa Streams" (*Des Moines Register* 2009).

Peter Gleick, one of the world's foremost water experts, reports that perceptions of water security are incorrect: over-pumped aquifers under breadbasket regions, such as the Ogallala Aquifer in the western United States, threaten agriculture, wasteful irrigation practices continue to deplete stream and river-based water systems, and water supply infrastructure is in a state of disrepair in urban centers worldwide (Gleick et al. 2006). According to businessman T. Boone Pickens, water has become "the new oil" (Berfield 2008).

Predictions of increasing water scarcity are usually driven by growing human populations running up against fixed supplies. The title of Jeffery Sachs' book,

Common Wealth: Economics for a Crowded Planet (2008), echoes this perspective which dates back to the Rev. Thomas Malthus' work in the seventeenth century. Like Malthus, Sachs hypothesizes that population growth will increase demand exponentially relative to fixed resource supplies. The inevitable result is that we will face resource shortages, including that of water.

Regarding what is often called "Blue Gold," Sachs argues that securing safe and plentiful global water supplies will prove to be one of our most daunting challenges. Sachs says that water scarcity is already a grim reality in many regions and that climate change will only further disrupt global water cycles, thus worsening the crisis (Sachs 2008, 115). This is because rural and urban societies around the world use more water than ever before, they use it at increasing rates, and they use it without regard for the future consequences of their consumption pattern (Sachs 2008, 121). Sachs further warns that privatizing water rights may be contrary to sound ecological management, especially for already overdrawn groundwater basins (Sachs 2008, 116).

To be sure, global population growth has increased water use substantially. Over the course of the twentieth century, most worldwide population estimates tripled while water use estimates increased more than sixfold. The world not only has more water users than a century ago, but each user is consuming more and more water (Cosgrove and Rijsberman 2000, 26). In *The Skeptical Environmentalist* (2001), Bjørn Lomborg reports that, per person, we have gone from using about 1,000 liters per day to almost 2,000 liters per day over the past 100 years[2] (Lomborg 2001, 151). By 2025, water withdrawals are predicted to increase from current levels by 50 percent in developing countries and 18 percent in developed countries (Zimmerman et al. 2008, 4251). Moreover, approximately 50 percent of precipitation recharges to groundwater in a natural system, whereas only 15 percent recharges in a highly urbanized environment, suggesting that population growth impacts water supplies in more subtle ways than simply increased withdrawals.

Access to Water

A closer look at the issue reveals that the primary concern is not absolute water supply but limited water access. In contrast to the inventory-based Malthusian theory of water shortages, water access relates to the distribution of water relative to human populations and the infrastructure that allows for water delivery, water storage, and basic sanitation. The statistics on water access are troubling. Though the planet holds approximately 326 million-trillion gallons of water, less than 0.4 percent is available and suitable for human consumption. The remaining 99.6 percent is either in the wrong place, the wrong form (such as ice or water vapor), or otherwise unfit for human consumption.[3] By World Bank (2003, 3) estimates, more than one billion people lack access to safe drinking water. Lomborg (2001, 154) reports that, though water accessibility has been getting better, there are still widespread shortages and limitations of basic services such as access to clean drinking water, and that local and regional scarcities occur.

The distribution of water supplies around the world simply does not match the distribution of people. North and Central America have 15 percent of the world's water, and only 8 percent of the world population. In contrast, Asia has 60 percent of the world's population but only 36 percent of the world's water. South America has 26 percent of the world's water and 6 percent of the world's population. Africa and Europe each have 13 percent of the world's population, and 11 percent and 8 percent of the world's water, respectively (Zimmerman et al. 2008, 4248). Given this asymmetry in water distribution and population density, almost a quarter of the world's nations currently lack sufficient fresh water to meet the needs of their burgeoning populations.

Most of the water-scarce countries are located in the Middle East and Africa, and most are poor. According to the United Nations Environment Programme, "[i]n Africa alone, it is estimated that 25 countries will be experiencing water stress[4] by 2025" (UNEP 2008). Water shortages in these areas threaten the health of humans and riparian ecosystems and hamper economic development. It is projected that by 2025, the Middle East, India, and most of North Africa will be withdrawing well over 40 percent of their total available water supplies, compared to less than 10 percent withdrawal of total supplies in Central Africa, South America, Australia, Canada, and Southeast Asia, and 20–40 percent in the United States, Russia, and Western Europe (UNEP 2007, 4252).

A look at the geography of water reveals that freshwater resources are often located at great distances from population centers and unevenly distributed. While there are approximately 263 river basins worldwide, some of the largest run through areas that are thinly populated (UNEP 2008). The Great Lakes of the United States, usable only by parts of the United States and Canada, comprise 20 percent of all surface fresh water, and Lake Baikal in southern Siberia holds another 20 percent. Rivers around the world hold only about 0.006 percent of total freshwater reserves (USGS 2009). According to the United Nations Development Programme, "[s]ome 1.4 billion people live in river basins where water use exceeds recharge rates." In these areas, "the symptoms of overuse are disturbingly clear: rivers are drying up, groundwater tables are falling and water-based ecosystems are being rapidly degraded" (UNDP 2006, 6).

Lack of access to clean water and sanitation has catastrophic effects on human health and productivity. Current estimates are that poor water quality, sanitation, and hygiene result in 1.7 million deaths per year internationally (Craft et al. 2006). Water-borne diseases are the greatest cause of infant mortality around the world. According to the World Water Council (2006, 224), "[i]n Central and Eastern Europe and Western Asia, it is estimated that greater than five percent of all childhood deaths are attributable to diarrheal disease, which is often a result of poor-quality drinking water, inadequate sanitation, or improper personal hygiene."

There is some evidence that water access is improving, but it still poses a significant threat to human health and welfare in most developing countries. Access to improved water supply increased 3 percent globally and 5 percent across Africa, Asia, Latin America and the Caribbean during the 1990s, while access to sanitation increased 5 percent globally and 9 percent across Africa, Asia, Latin American and the Caribbean during the same period (WHO and United Nations Children's Fund 2000).

These improvements in water access are significant and worth celebrating, but water scarcity poses a threat to human health that cannot be eradicated. Mismanagement and inefficient allocation can quickly reverse improvements in water access and sanitation, as could population growth, political instability, and climate change. No water source is inexhaustible, as we are observing throughout the world.

Changing Sources of Water Withdrawal

Related to the issue of water access are changes in the source of water withdrawals, particularly from surface to groundwater. According to Alley et al. (2002),

> during the past 50 years, groundwater depletion has spread from isolated pockets to large areas in many countries throughout the world. Prominent examples include the High Plains of the central United States, where more than half the groundwater in storage has been depleted in some areas, and the North China Plain, where depletion of shallow aquifers is forcing development of deep, slowly replenished aquifers with wells now reaching more than 1000 meters.

Villagers in the northern Hebei province of China are digging wells 120 to 200 meters deep to find clean drinking water where wells were only 20 meters deep a decade ago (Gleick et al. 2009, 86). In southern India, the 21 million wells drilled are lowering water tables in most of the country—in Tamil Nadu, for example, the falling water table has dried up 95 percent of wells owned by small farmers. A similar story exists in northern Indian states such as Gujarat, where the water table is purportedly falling by 6 meters every year (Pearce 2006).

The problems created by excessive groundwater withdrawals are neither recent nor confined to developing countries. In sections of Kansas and Texas, the Ogallala Aquifer had dropped 150 feet by 1980 (Glennon 2002, 26). The United States Geological Service estimates water in storage in these parts of the aquifer has declined approximately 190 million acre-feet[5], a 34 percent decline since large-scale irrigation began (McGuire et al. 2003). As a result, groundwater depletion may be the single largest threat to irrigated agriculture worldwide.

Water Pollution

Pollution of both surface and groundwater poses an additional threat to future water supplies and access. In 1975, the U.S. Water Resources Council was optimistic that "water-quality conditions will be improved substantially. With the emphasis on more intensive use and reuse of available supplies, improvement of quality should become an important facet of water-management procedures" (U.S. Water Resources Council 1978, 78). Unfortunately, that optimism has not blossomed into reality.

The World Resources Institute (2008) has identified 415 eutrophic and hypoxic coastal systems worldwide.[6] Of these, 169 are documented hypoxic areas and 233 are areas of concern, while only 13 are systems in recovery. In the United States, 78 percent of the continental coastal area and 65 percent of the Atlantic coast are eutrophic.[7] In Asia, only 35 percent of wastewater is treated, compared to only 14 percent in Latin America and less than 1 percent in Africa.

In the United States, the Great Lakes remain plagued by phosphates from household laundry detergent, toxic chemicals discharged by industry, pesticides that have drifted on air currents and settled in the lakes, and fertilizer runoff. The Chesapeake Bay receives toxic industrial wastes, harmful substances from solid waste, and pesticides from farms and other non-point sources. Both public and private efforts have been undertaken to restore the quality of these water bodies, but little progress has been made.

The Potential for Water Markets

The regulatory response to water scarcity is to restrict withdrawals and consumption. Low-flow technology mandates, water rationing, and use restrictions can alleviate short-term and small-scale water shortages through forced reductions in demand. Such regulatory responses ignore the economic forces that produce and perpetuate worldwide water scarcity.

This book explores an alternative approach to solving the world's water problems—water markets. In the most fundamental sense, water markets are a voluntary exchange between willing buyers and willing sellers of legal rights in water. Water markets can be informal trading arrangements between ranching families or inter-basin transfers of enormous volumes across hundreds of miles. The important and distinguishing features of all competitive water markets are (1) the voluntariness of the trade and (2) the legality of the right traded. Markets, whether as simple as a trade between school children on the playground or as complex as international derivative markets, effectively move goods from lower to higher valued uses, utilizing prices, contracts and self-interest to do so. Rarely does one hear of a crisis in condominiums, pick-up trucks or laptop computers, mainly because markets work to eliminate shortages by balancing demand and supply.

Yet opposition abounds to using markets to allocate scarce water resources. Some consider water sacred and a human right (Barlow 2007). Others argue that water is a basic human necessity that should not be withheld from those who cannot pay, be supplied by profit-seeking corporations, or otherwise left to the vagaries of competitive markets (Dosi and Easter 2003). The thrust of these arguments is that, at least in principle, water and markets are incompatible.

Ours is a world of water scarcity, however, so conservation and allocation matter more than claims of right and water sanctity. In this book, we contend that because water is so important, markets are a too-little used tool for allocating scarce water resources. Without markets and prices to provide incentives for both demanders and suppliers, water crises will persist.

Two examples illustrate how water markets can be thwarted or encouraged, both of which involve the complex problem of interstate water allocation. In Wisconsin, the Great Lakes Basin Water Compact disallows water trading outside the compact area, thus creating a water crisis for the town of Waukesha. In Colorado, environmentalists, ranchers, and municipalities are developing innovative market solutions to reduce the effect of other states in the basin calling for more water, thus preventing a crisis. What is the difference between the two?

Waukesha, Wisconsin

This small town just 20 miles west of Lake Michigan has a problem that would surprise most folks from outside the area: it is running out of water. Despite the region's heavy precipitation and the community's proximity to one of the largest freshwater supplies in the world, the Waukesha water utility cannot find supplies to meet projected demand. Water scarcity in a humid place like Wisconsin has long been rare, but is becoming more common. Hence, the history and future of Waukesha's water troubles has broad implications for the Midwest and beyond.

The hydrology of the region explains part of the problem. The deep sandstone aquifer underlying southeastern Wisconsin and northern Illinois is highly confined by subterranean features like the Maquoketa shale (Jansen and Schultz 2009). These features enclose the aquifer and prevent it from recharging quickly. Indeed, hydrologists estimate the aquifer draws only 1.3 million gallons per day directly from Lake Michigan (U.S. EPA and the Government of Canada 2002). That amount may sound like a lot, but it's not. On average, Lake Michigan loses that much water via surface evaporation every five seconds!

Given this hydrological isolation, decades of pumping by Waukesha and other communities have created a cone of depression 500 to 600 feet below pre-development levels (Barrett 2009). The water table's decline means pumping is more difficult and costly than ever before. Adding to the problem, radium and sodium levels in the water have steadily increased with pumping pressures. The U.S. Environmental Protection Agency regulates naturally occurring radioactive elements and has required, along with the state of Wisconsin, that Waukesha reach full compliance with radium concentration standards by 2018—a requirement that could cost millions in treatment costs. In short, groundwater is becoming very expensive in Waukesha.

The second reason Waukesha faces a water crisis has more to do with politics than hydrology. The Great Lakes Basin Compact prevents any new diversions of surface water outside the Great Lakes Basin. Because portions of Waukesha County lie outside the sub-continental divide (the line between the Great Lakes and Mississippi River basins), the prohibition prevents Waukesha from diverting water from Lake Michigan without first obtaining approval from each of the eight states and two Canadian provinces that signed the compact.[8] Any single signatory to the compact may veto the application, which means the Waukesha water utility has a high hurdle to clear before it can access Lake Michigan water.

If any application had a chance, the following scenario would be it. The city of Milwaukee has excess capacity in its water system and could sell its water to Waukesha for approximately $2 million in annual water fees. In return, Waukesha would receive much needed water from Lake Michigan, a less depleted water source than the over-pumped aquifer it currently taps. As the proposed plan includes a $78 million dollar pipeline construction project, a veto of the deal by one of the compact states would also have broad economic implications for the region. From both an economic and environmental perspective, it's a deal that would make a lot of sense.

However, the odds are probably low that Waukesha's application will be approved by every signatory to the compact for one simple reason: this is the first application for a new diversion outside the Great Lakes Basin and, by implication, the first challenge to the Great Lakes Basin Compact. As such, it sets an important precedent for future applications and other legal challenges. If Waukesha's application is approved, the compact signatories can expect a flood of exemption applications from similarly situated communities all over the region. Despite Waukesha's proximity to Lake Michigan, the compact creates miles of red tape that entangle the potential for trade.

The Colorado Front Range

Unlike the Great Lakes region, water scarcity is common throughout the Rocky Mountain West and has been since frontier days. Scarcity and the potential for conflict, however, are increasing each day as a result of population growth. Between 2000 and 2010, the Rocky Mountain West boasted the fastest growing population of any region in the country (Mackun and Wilson 2011). Consequently, municipalities from Arizona to Montana are struggling to find new water supplies sufficient to meet the demands of growing urban populations.

Cities and towns along Colorado's Front Range have experienced even faster growth than other parts of the region. Between 2000 and 2008, for example, the population of Fort Collins grew by 16.43 percent, Greeley's by 38.05 percent and Douglas County's (midway between Denver and Colorado Springs) by a staggering 59.66 percent (U.S. Census Bureau 2009). Population growth of this magnitude strains municipal water supplies and infrastructure, exacerbating the water stress caused by Colorado's prolonged drought.

In addition to the hydrological constraints on water consumption, an interstate agreement similar to the Great Lakes Basin Compact may force the Front Range municipalities to cut back even further. The Colorado River Compact, put into effect in 1922, adjudicates the distribution of water among the seven states in the Colorado River Basin. The compact prohibits the Upper Basin states (Colorado, New Mexico, Utah, and Wyoming) from depleting the river's flow below 75 million acre-feet for any period of 10 consecutive years. Stated differently, the Upper Basin states cannot reduce the 10-year rolling average below 7.5 million acre-feet.

Because of persistent drought and increased diversions, less and less has been flowing from the Upper Basin to the Lower Basin (Western Water Assessment 2011). If the 10-year rolling average falls below 7.5 million acre-feet, the Lower Basin states (Arizona, California, and Nevada) may institute a forced reduction in Upper Basin water consumption in what is known as a "compact curtailment." Western water law requires that curtailment begins with the most junior rights holders—those who claimed water most recently—and progresses to more senior rights holders until the level of curtailment is met.[9] Under the compact, rights established after 1922 when the compact was signed are junior rights that would have to be curtailed until the minimum 10-year rolling average is achieved. Senior water rights or those perfected before 1922, on the other hand, would be unaffected by the curtailment.

Irrigators on Colorado's West Slope hold most of the state's senior water rights, while Front Range municipalities hold primarily junior water rights—rights that would be cut in the event of a compact curtailment. The prospect of a compact curtailment therefore threatens the ability of Front Range municipalities to provide basic water and sanitation services. As a consequence, government officials are scrambling to find more secure water supplies.

One possible solution under consideration is a water bank that would allow Front Range municipalities to trade water diversion rights with West Slope irrigators. The water banking proposal would allow senior rights holders to deposit their rights into a bank from which junior rights holders could lease or purchase rights instead of curtailing their consumption. Thus, if the 10-year rolling average falls below the 7.5 million acre-feet minimum and a compact curtailment is required, municipalities could pay senior right irrigators to forego their diversions so that junior right municipalities could continue serving domestic consumers.

The critical element of the banking concept is that West Slope irrigators putting water to a lower valued use would be compensated for leaving their water instream so that Front Range municipalities with higher valued uses will not have to curtail consumption. In this way, water bank traders could ensure that consumption rights flow to higher valued uses while also ensuring that Colorado complies with the compact. Because irrigators and municipalities would only trade the seniority of their water rights, no physical transfer of water is necessary.

Environmental organizations support the water banking concept because it provides an opportunity for irrigators to leave more water instream for fish and fish habitat. Indeed, organizations such as Trout Unlimited or the Nature Conservancy could use the bank to purchase senior diversion rights and leave water instream.

The proposed water bank exemplifies the flexibility of water marketing. No physical transfer of water is required, participation is voluntary, and no government bureaucracy is necessary.

Looking Forward

The remainder of this book explains how water markets can ameliorate water crises by balancing demand and supply. Chapter 2 contrasts market solutions with regulatory solutions. Chapter 3 reviews the history of water institutions in the United States and shows how they have evolved in response to water scarcity. Chapter 4 explains how politics distorts market decisions, especially when it subsidizes massive storage and delivery systems. Chapters 5 and 6 outline the necessary conditions for water markets to function in prior appropriation and riparian states, respectively. Chapters 7, 8, and 9 analyze specific water issues—instream flows, water quality, and groundwater, respectively. The book concludes with an outline of the most pressing issues related to domestic and international water markets and argues that water markets can play an integral role in keeping the Blue Planet blue.

2

CHEAPER THAN DIRT

The predictions of water crises outlined in Chapter 1 are intuitively appealing, especially in light of a growing global population with ever-increasing water demands and a finite supply of water found on the Blue Planet. With such an imbalance between the quantity people demand and the quantity that is supplied, water crises seem inevitable and potentially catastrophic. In his aptly named book, *Collapse* (2005), famed geography professor Jared Diamond carefully details how water stress contributed to the collapse of several ancient societies and how the conditions leading up to those societal collapses are present throughout the modern world. This link between societal collapse and water scarcity gives new meaning to the quip attributed to Mark Twain that "whiskey is for drinking; water is for fighting over."

The extent to which predictions of water crises warrant concern depends on whether there are institutions that balance the quantity of water demanded with the quantity supplied. One such institution of growing importance is water markets, from which come water prices, the signaling device that can correct the imbalance between the quantities demanded and supplied. Indeed, the economic law of demand applies as much to water as it does to any other good. Higher prices will lower the quantity demanded and lower prices will increase it. To be sure, there is some level of water consumption below which any species, human or other, cannot survive, which underscores the importance of institutions that signal water's scarcity and value.

A well-functioning market for water has the potential to turn crises into cooperation when willing buyers and willing sellers strike deals that balance water demands with water supplies. Without detailing until later the conditions necessary for well-functioning water markets, suffice it to say here that numerous examples suggest that markets can play an important role. In watersheds throughout the American West, for example, irrigators lease water to environmentalists for instream flows. These deals benefit both parties involved and, they reflect the ability of markets to adjust to changing water demands. Between 1987 and 2007, $530 million (to be sure much from governmental agencies) was spent to purchase over 10 million acre-feet of water for instream flows from willing sellers, mainly farmers and ranchers (see Scarborough 2010).

Contrast the ongoing battle between environmentalists and farmers over water in California's Central Valley with the potential for cooperation between similar groups in Oregon. In California, the conflict is over millions of gallons of water in the San Joaquin delta. Central Valley farmers want the water for irrigation, southern California cities want the water for municipal use, and environmentalists want the water to stay in the delta for the threatened delta smelt and salmon species. Because the water is typically delivered to farmers through state and federal water projects at subsidized rates, the question of who gets the water is really a question of who has the most political clout. As that pendulum of political power swings back and forth, neither the fish nor the farms are safe.

In Oregon, environmentalists and farmers are using markets to cooperatively and effectively transfer water from agricultural diversions for irrigation to stream flows for salmon. In 2006, for example, the Oregon Water Trust (OWT) (now the Freshwater Trust after a merger between OWT and Oregon Trout in 2009) entered into an agreement with a third-generation ranching family, Pat and Hedy Voigt, who agreed to permanently shorten their irrigation season. By purchasing some of the Voigt's rights to divert water, OWT was able to keep 6.5 million gallons per day in the John Day River in the late summer when the last and largest remaining populations of spring Chinook salmon and summer steelhead need it. From its start in 1994 with only two leases totaling 1.4 cubic feet per second, the trust has converted nearly 53,000 acre-feet from diversions to instream flows thus benefitting nearly 900 miles of streams. Its projects represent cooperation with more than 200 landowners, one of whom said, "I give OWT a lot of credit. They came to us with an attitude of wanting to help and they displayed a great deal of respect for agriculture" (quoted in Scarborough and Lund 2007, 51).

The contrast between the political battles in California where massive, subsidized water projects allocate water and the cooperation in Oregon where the Freshwater Trust is trading with farmers illustrates how different the two processes are. Water markets allow competing water users to find gains from trade by increasing water-use efficiency. Though wringing hands and gnashing teeth over water crises may make good headlines and fodder for political debates, these strategies do little to actually solve water scarcity problems.

In this chapter, we compare and contrast the potential for water markets to resolve water crises with regulatory regimes that allocate water and set prices through political processes. Traditional regulatory approaches to water policy have led to subsidies, price distortions, and environmental destruction. In contrast, water markets have and could do even more to promote fiscal and environmental responsibility. The latter emphasizes the importance of getting the incentives right, something that is very difficult to accomplish through politically charged water regulations. The basic conclusion is that while markets may not work perfectly, they can be an improvement over existing regulatory policies, especially with regard to water-use efficiency, environmental quality, and fiscal responsibility.

Markets vs. Politics

Economic policy analysis emphasizes efficiency and the potential for maximizing social welfare by ensuring that the marginal social benefits of resource use are equal to the marginal social costs, but it does so mostly in an institutional vacuum. Accordingly, if an additional unit of water is worth $10 to industry and worth only $9 to agriculture, it will be efficient and welfare enhancing to reallocate water from lower valued agricultural uses to the higher valued industrial uses. Under normal assumptions about demand and supply, such reallocation will increase the marginal value of water left in agriculture and decrease the marginal value of water to industry, thus equilibrating marginal benefits and marginal costs.

This simple yet powerful principle explains what is meant by efficient water allocation, but it does not explain how markets or political institutions (legislatures, agencies, or courts) account for marginal costs and benefits or, as a consequence, whether market allocation or political allocation is more likely to promote efficiency.

Perhaps the biggest difference between market and political allocation is in the way that the value of water is determined. Market exchanges allow buyers and sellers to weigh their private marginal costs with their private marginal benefits and to express their values through prices usually, though not necessarily, expressed in money terms. If there are gains from trade, the buyers and sellers can potentially bargain and agree on a price that makes them both better off, and they can continue to do so until all the gains from trade are exhausted; in other words, until marginal benefits equal marginal costs and an efficient allocation is attained. Assuming that the costs they face and the benefits they receive capture the values to society (an assumption we will examine thoroughly in chapters to come), social welfare will be maximized.

In contrast, political allocation relies on experts who measure costs and benefits with a different scale. They might include market values observed in other markets or obtained from economic cost–benefit analyses (as opposed to prices bargained through trade). More importantly, they will rely on political signals provided from voting, lobbying, interest group pressures, hearings, and polls, to mention a few. The question, therefore, in comparing water market allocation with political allocation is whether a voting majority, arguments by lobbyists, pressure from interest groups, and so on reflect the social values better than market exchanges. To be sure, a lesson from public choice economics is that private opportunity costs will play less of a role in political decisions (see Mitchell and Simmons 1994).

Given the very different processes and scales used by market participants versus political allocators, it is not surprising that water allocation is quite different under these alternative processes. A more thorough analysis of each approach shows how they affect water allocation.

Do Price Signals Work?

Prices are an indispensable part of the producer's or consumer's cost–benefit calculus. When scarcity drives up the price of a resource, there is a reward to resource users who find alternative sources of supply, new technologies, or substitute resources. Indeed, one economic study after another[1] shows that price responses have played a crucial role in helping humankind avoid "the limits to growth" predicted by the Reverend Thomas Malthus in the seventeenth century and carried on by modern-day Malthusians such as the Club of Rome (Meadows et al. 1972) and Paul and Anne Ehrlich (Ehrlich 1968) and espoused in the *Global 2000 Report* (Barney 1980). Whether it is new resource discoveries or technological changes on the supply side or reduced consumption on the demand side, prices are a driving force behind human responses to averting crises.

Before turning to the price responsiveness of water demand and supply, it is important to understand the link between a resource crisis and the price of the resource. Consider, for example, the energy crisis of the late 1970s and early 1980s. Prices of energy had fallen steadily since World War II despite the growing demand for it to power our transportation—"See the USA in a Chevrolet"—and our homes—"Live better electrically." Then in the early 1970s, OPEC restricted oil production and drove up its price at a time when the U.S. government had imposed price controls to deal with inflation. The energy crisis resulted because consumers, accustomed to high energy consumption, could not easily adjust—gas guzzlers continued to guzzle and home insulation remained thin—and producers could not raise their prices. In economic parlance, the quantity demanded exceeded the quantity supplied, and the price could not rise to balance the two. Instead, lines formed at gas stations, rationing was common, and people marched in the streets to protest the crisis. In contrast, when energy prices rose dramatically in 2008, drivers drove less and conserved more while producers increased supplies from existing sources and searched for new ones (Keen 2008). When prices fell back in 2009, total gasoline consumption increased while investment in developing new supplies fell. In stark contrast to the energy crisis of the 1970s, the dramatic price fluctuations in the 2000s were taken in stride by both consumers and producers.

In economic terms, therefore, resource crises are the result of prices not balancing the quantity demanded by consumers with the quantity supplied by suppliers, and water crises are no different. Low prices encourage a high quantity of water demanded by consumers and a low quantity supplied by producers. Water prices today are seldom set by market forces but are often fixed, so real water price signals are lacking. However, where they do exist, efficiency and conservation gains from marginal adjustments are significant. For example, a comprehensive analysis of agricultural water demand studies conducted between 1963 and 2004 found that, on average, a 10 percent increase in the price of water would reduce demand by nearly 5 percent (Scheierling et al. 2006). The quantity of water demanded by industry in France declines on average by 1–7.9 percent, depending on the industry, for a 10 percent increase in water rates (Reynaud 2003). The long-term effects can be even greater as agricultural and industrial water users employ new technologies and capital improvements to improve water-use efficiencies.

The quantity demanded by households similarly declines in response to water price increases. In the short term, the quantity demanded declines by roughly 3–4 percent for every 10 percent increase in price, while the long-term quantity demanded falls by a full 6 percent (Espey et al. 2007). These reductions reflect both short- and long-term responses to rising water prices. In the short term, consumers reduce water bills by reducing usage. In the long run, they move to new technologies to improve water-use efficiencies. For example, consumers may replace outdated fixtures and appliances with more efficient ones, or change landscaping from water-intensive vegetation to xeriscaping techniques (Olmstead and Stavins 2007). The implications of the price responsiveness of consumers are significant. If subsidies keep water prices artificially low, an imbalance between demand and supply will occur and alternative forms of rationing—lawn watering restrictions, mandated low-flow toilets, limits on new housing starts—will have to be used.

Low prices encourage all consumers, municipal, industrial, agricultural, and environmental, to march down their demand curves and use the increasing amounts of water in uses with low marginal values. Therefore what is seen as waste or inefficient water use in agriculture—irrigation water eroding a field without reaching the roots of the crops—and domestic use—water for lawns running down the storm gutters—is simply the user's rational response to low water prices. Even environmental uses have decreasing marginal value. Enough water to keep fish alive is less than enough to provide long-term spawning habitat, which is less than enough to keep a stream running bank full. What appears to be wasteful or inefficient use is a reflection of low prices that allow low valued water uses. To the consumer, they are simply a rational response to cheap water.

The increased use of water as it becomes cheaper is particularly evident in agriculture. If water were more expensive, less would be applied to any given crop, different irrigation technology or water application practices would be used, and different cropping patterns might appear. Research conducted at the University of California shows that reduced water application would decrease most crop yields but that at higher water prices, such reductions would be economical (Caswell and Zilberman 1985). Flood irrigation techniques conserve on labor but require large amounts of water. With high water prices, it makes sense to substitute labor and capital for water by switching to drip irrigation or similar techniques. Trimble Hedges (1977) provided similar evidence in a simulation of a 640-acre farm in Yolo County, California. Hedges showed that the optimal cropping pattern at a zero water price would call for 150 acres each of tomatoes, sugar beets, and wheat; 47 acres of alfalfa; 65 acres of beans; and 38 acres of safflower. If the water price were increased to $13.50 per acre-foot, alfalfa acreage would drop out and safflower acreage, a crop that uses less water, would expand. The point is that many choices are available to water consumers, and they will respond rationally to market prices for water.

Because water prices motivate agricultural users to reduce their consumption by using improved irrigation techniques and modified cropping patterns, markets have the potential to free up irrigation water for municipal, industrial, and environmental uses which often have higher marginal values. For example, water withdrawals for irrigation are roughly 128,000 million gallons per day. Trading 5 percent of that water to municipalities would supply sufficient water to an additional 64 million residents (assuming per capita use of 100 gallons per day) (Kenny et al. 2009). However, for such trading to occur, agricultural users must bear the full cost of their water consumption, including the opportunity cost of not transferring water to other users.

The Potential for Water Markets

Water markets have great potential for generating prices that reflect water scarcity and for switching on incentives for innovation. Individuals, especially entrepreneurs, and firms are the starting point for market transactions (Anderson and McCormick 2004). Motivated by profits, entrepreneurs search for opportunities to move resources from lower valued to higher valued uses. As they respond to such opportunities, resource allocation is improved, not because the entrepreneur cares about efficiency, but because profits in a competitive market are achieved by pursuing efficiency.

If entrepreneurs face the full costs of their actions, they will take only those actions that produce positive net benefits. As long as the entrepreneur who discovers a higher valued use for water and transfers water to that use bears the opportunity costs of current use, the reallocation will only take place if the net difference is positive. Responsibility for opportunity costs is crucial. Suppose, for example, that a corn farmer sells his irrigation water to an electrical generating plant that uses the water for cooling. In the process, assume that less water is actually consumed, i.e., less is removed from the source and stored in the corn kernels, but that the temperature of water discharged is higher. Will the trade generate positive net benefits? Obviously the generating plant will have to pay the opportunity cost of the water in corn production or the farmer would not be willing to sell. But what about the values of water returned to the stream? More water in the stream may have positive value for some people, say river rafters, but warmer water may have negative values, say for trout fishers, if it decreases trout populations. If the latter two are not accounted for in the transaction, we cannot be sure of the net value of the transfer of water from corn to electricity production.

Well-defined, enforced, and tradable property rights are key to getting owners to take account of opportunity costs. Consider each of these three elements in the context of water rights. Defining water rights requires some unit of measurement. In the early mining camps, rights were specified in "miner's inches," the amount of water that would flow through a one-square inch hole cut in a board inserted in the channel. Today it is more common to measure flows in gallons or cubic meters per second and volumes in acre-feet—the amount of water necessary to cover one acre of land one foot deep with water. Well-defined water rights must specify the quantity of water that can be diverted from a stream, the timing of the diversion, and the quantity and quality that must be returned. When not all claims on a stream can be met, water rights must specify whose rights, if any, have priority or whether all must reduce diversions proportionately.

Well-enforced water rights ensure that an owner can enjoy the benefits of her ownership without those benefits being taken by others. This means that there must be some way of monitoring water stocks and flows. Enforcement allows owners to exclude other users and therefore to capture the benefits from the uses of their water unless they are compensated to give up those benefits. If rights are not well enforced, on the other hand, others get to use the water without paying.

Of course, ownership is always probabilistic. How well defined and enforced water rights are depends on how much effort owners put into definition and enforcement (see Demsetz 1967; Anderson and Hill 1975). As water rights evolved on the western frontier, little effort was put into definition and enforcement until one diversion reduced the potential for another. Then miners met in their camps and farmers formed ditch companies to establish rights for diversion. Technology limited measurement to boards inserted into channels or measuring sticks attached to diversion structures. Recordation was informal and records were often lost. Today, in contrast, instream sensors, hand-held radar, and hydroacoustic devices that continuously monitor and record stream flows can provide accurate flow and volume measurements, often in real time. And to overcome poor recordation, some states have undertaken long and expensive adjudication processes to clarify rights, reflecting the fact that as water has become more valuable, it is worth more for claimants to invest in local measurement and monitoring and centralized recordation and enforcement.

If water rights are not well defined and enforced and ownership claims are weak, stewardship and conservation are unlikely. For example, if a water user does not have the right to use or sell water he conserves by installing a more efficient irrigation system, he will have little incentive to improve water-use efficiency. Or if a water owner decides to leave water in the stream to improve fish habitat but others are free to divert it for irrigation or if anglers can fish in the stream without paying, he will be less likely to enhance stream flows.

Finally, water rights must be transferable if the owner is to be fully aware of the opportunity costs of his actions. If a water owner is not allowed to transfer his water to a higher valued use or user, he will ignore the opportunity costs of the current use and ignore the increased value that could be achieved. Laws forbidding the sale or lease of diverted water to environmental groups for enhancing environmental amenities, for example, tell irrigators to ignore environmental values.

How Pervasive Is Market Failure?

Because property rights are never perfectly defined, secure, and transferable and because of the potential for third-party effects and free riding, markets may not accurately reflect the full value and cost of water use. George A. Gould (1988, 22) states that "[t]hird-party effects greatly impede the development of markets in appropriative [water] rights." More recently, Jeffrey Sachs (2008, 115) claimed that "[p]rivatization of water rights may be contrary to basic ecological good management [because] water is characterized (appropriately enough) by pervasive spillover effects." If third-party effects are not accounted for by the legal system— in other words, if property rights are not clearly defined and enforced— governmental intervention and regulation may be justified.

These arguments imply that individuals not involved in a market transaction and whose rights are violated bear some costs not accounted for by the trading parties.[2] If so, the harm would have to be weighed against the gains from trade, potentially making the net value of a transaction zero or negative.

For example, suppose that one water user who values water at $20 per unit purchases water from another who values it at $15. The potential gain from trade appears to be $5. However, if a third party's rights are violated, say because water quality declines as a result of the new use in the amount of $10, there is actually a net loss when all costs are accounted for. Of course, if the harmed party has a property right to clean water and can make the others pay for the harm, no exchange between the first two parties would take place because the gains from trade are not sufficient to compensate for the harm.

Third-party effects occur when resources are held in common with access available to everyone. For example, pumping by one person from a groundwater basin can result in higher pumping costs for others who pump when the water table is lower. This is the essence of the tragedy of the commons explained in Garrett Hardin's seminal article by that title (1968). While it is true that individual pumpers extracting water from a basin do not take account of their effect on the pumping costs of others, the relevant question is why do not these others extract a payment for the harm caused? As A. C. Pigou, considered the godfather of market failure arguments, put it, market failure might occur if "compensation [cannot be] enforced on behalf of the injured parties" (Pigou 1932, 183). The inability to enforce compensation would happen due to what Pigou called a "technical difficulty," or what Nobel laureate Ronald Coase (1960) called transaction costs.

The existence of technical difficulties or transaction costs requires focusing on what these difficulties or costs are. Do the newcomers to the groundwater basin have a right to lower pumping costs? To be sure the costs are higher due to the first pumper, but if latecomers do not have a right to lower pumping costs, they cannot claim harm. And if they do have such a right, why can they not enforce their right by calling on the state legal system to enforce a payment for the harm? Is it because their rights are not clearly specified or is it too costly to enforce the rights relative to the benefits of so doing? Suffice it to say transaction costs are real costs of markets in the same way that transportation costs limit whether goods will be traded across vast distances (see Demsetz 2003). The question that needs to be answered before water markets are judged to fail is why the legal institutions are not enforcing the rights. In the following chapters, we will consider this question and provide examples where clarifying property rights and allowing trade is overcoming third-party effects.

Another potential cause of market failure is the cost of information. For traders to exhaust the gains from trade and improve water-use efficiency, they must know what the benefits and costs are and, especially with a resource such as water, this knowledge comes with uncertainty. For example, weather patterns are stochastic, making it difficult to predict how much water there will be available for irrigation or for fish habitat. And with the current predictions of climate change, whether caused by anthropogenic activities or not, forecasting long-term trends is even harder. In the same way transaction costs are an alleged cause of market failure, information costs are an inherent cost of resource allocation. Again, the question is whether markets or political institutions can better cope with these costs. Market exchanges require individual traders to make their best calculation of uncertainty and to utilize market tools such as futures contracts, option contracts, and derivatives, to mention a few, as a way of adjusting to risk and uncertainty.

To illustrate the potential for water markets, consider a contract between OWT and an irrigator in the Walla Walla Basin. The trust recognized that it only needed to reduce diversions and increase instream flows when flows were low, for example, as the result of a drought. Of course, drought conditions also make the water more valuable for irrigation. Neither party could perfectly predict when a drought might occur, and OWT did not want to pay for water when it was naturally abundant. By offering the wheat and pea farmer an upfront payment for the option to lease his water when stream flows dropped below a certain level, OWT was able to reduce its costs while still paying the farmer for foregone production. Of course, there were still information and uncertainty costs, but the parties had an incentive to minimize them within their constraints.

Gathering information with which to make decisions requires a careful comparison of the benefits and costs of the information itself. Should a farmer install a new sprinkler irrigation system that reduces water use by 25 percent? What is the cost of the capital? What is the value of the water saved and can that value be captured? How long will the system last? There is an endless list of questions that can be asked but, at some point, the value of additional information compared to the cost of acquiring that information makes further study uneconomical. Reducing uncertainty is good but only if the expected benefits from the search activity exceed the expected costs. Of course, what may be the optimal amount of search for one person or one situation will not necessarily be optimal for another, so it is easy for observers to argue that better decisions could result if more information were collected. But perfect information is not the norm to which we should compare the real world and therefore should not be used to form the conclusion that markets fail. Market decisions, like political ones, require gathering and acting upon information relevant to the decisionmakers. As we shall see in the next section, that information and the action that results differ significantly between the two.

A final argument regarding the failure of water markets focuses on equity rather than efficiency. The leading proponent of this argument is Maude Barlow, who argues that "life requires access to clean water; to deny the right to water is to deny the right to life" (2007, xii). She warns that the commodification of water and its transfer on the open market will deepen the water crisis and spawn a "corporate water cartel [that seizes] control of every aspect of water for its own profit" (Barlow 2007, 2). As Hirshleifer et al. (1960, 367) recognized decades ago, such thinking leads to "the near-universal view that private ownership is unseemly or dangerous for a type of property so uniquely the common concern of all."

Granted, that some amount of water is a necessity of life and therefore that equity cannot be ignored in water policy, but it does not follow that markets have no place in water allocation. Barlow's concerns could easily be addressed by granting all humans a right to an amount of water necessary for consumption and hygiene, say 15 gallons per day, and letting anything more be allocated by markets. To declare that access to water is a human right but not address how that right will be met will ensure that supply always falls short of demand.

Can Politics Fix the Failures?

The extent to which water should be allocated by market processes or political processes requires a comparison of which approach better deals with the criticisms discussed above. To determine how water markets work relative to the political allocation of water requires an understanding of public choice economics, the branch of economics that incorporates economics and political science.[3] Public choice recognizes that the incentive structure in the public sector is quite different from that in the private sector. In the private sector, owners of firms receive profits equal to the difference between revenues and costs. Given this claim on the residual, the owner has an incentive to find good information and use it to improve efficiency, which in turn enhances profits. In the political sector, however, there is no owner with a claim on the profit. The rewards for state water engineers do not depend on maximizing the net value of water resources and the salaries of managers of Bureau of Reclamation projects do not depend on whether farmers pay off their loans. Votes, campaign finance, political influence, and budgets are all factors that affect decisions in the political sector, but there is little consensus in the economics literature as to just how these fit into the political and bureaucratic process. Making tradeoffs based on these political currencies, at best, distorts measures of economic value and distorts resource allocation away from efficiency. In short, efficiency and profits have no constituency in the political arena.

If politicians and bureaucrats are not motivated by efficiency and profits, they also are not scientific managers methodically applying cost–benefit analysis to achieve some clear vision of optimal water use. In his book, *Conservation and the Gospel of Efficiency*, Samuel P. Hays distinguished between the rhetoric of "a political system guided by the ideal of efficiency and dominated by the technicians who could best determine how to achieve it" and the reality of "a limited group of people, with a particular set of goals [whose] definition of the 'public interest' might well, and did, clash with other competing definitions" (Hays 1959, 3–4). Using examples from public land management, reclamation, and forestry, Hays demonstrated how the bureaucratic process can obfuscate cost–benefit comparisons and produce outcomes inconsistent with efficient resource management. Complete and accurate information is rarely available to the bureaucratic technicians intent on achieving efficiency.

Applying cost–benefit analysis to political decisions could help replicate the efficiency of market forces, but doing cost–benefit analysis is more of an art than a science and even when it is carried out properly, it is often ignored. Outputs produced in the public sector are almost never priced and, if they are, the prices do not necessarily reflect the costs. For example, the rates charged by water and waste water utilities, the vast majority of which are publicly owned and operated, often fail to cover the costs of providing those services. According to a 2002 report from the U.S. General Accounting Office, 29 percent of water utilities and 41 percent of waste water utilities are generating revenues from user rates that are insufficient to fully cover service costs. The reason for the funding shortfall, according to water expert Tracy Mehan (2008), is that elected officials avoid rate increases, particularly in election years, in order to gain political capital.

Accurately forecasting, discounting, and then weighing future costs and benefits are challenging for private sector firms, but the incentives of term-limited elected officials make performance of these tasks even less likely in the public sector. Consider again the routine underpricing of water and waste water services and the attendant consequences of deferred infrastructure maintenance and investment. The U.S. Environmental Protection Agency estimates the investment gap in capital needs for water and waste water utilities nationwide exceeds $220 billion (U.S. EPA 2002). Failure of these systems will leave countless communities without water or sanitation services and could generate rebuilding costs orders of magnitude larger than preventative maintenance costs, yet an elected official would rationally ignore these future costs if they are likely to accrue after his term expires.

Even if politicians and bureaucrats have good intentions, they still face the problem of knowing what the voting public wants, what goods and services are worth, and what they cost to produce. The economics of public choice teaches us to view public sector activities like any other activity. Politicians and bureaucrats are trying to maximize certain objectives such as votes, budgets, power, prestige, and discretion. As these goals are pursued through collective action, public decisionmakers face costs that differ substantially from those in the private sector because politics allows those who bear the costs to be separated from those who receive the benefits.

Building on the premise that actors in the political system are motivated by self-interest, the public choice paradigm has uncovered several reasons why governmental action has failed to meet efficiency criteria:

1. *Voter ignorance and imperfect information.* In a democratic society where it is unlikely that any voter, even a well-informed one, can influence the outcome of the political process, the benefits of being well informed cannot be fully

captured by the individual. At the same time, obtaining information about candidates and issues is costly for the voter. Thus, voters remain rationally ignorant; that is, they do not undertake great costs to obtain information except on issues that are important to them personally.

2. *Special interest effects.* Those voters who do become well informed and politically active on any issue tend to be those who will benefit from a particular governmental action. With benefits concentrated on a few recipients, it is worth the recipient group's time to try to bring about specific governmental action. Since the costs of governmental action—subsidies, transfer payments, tariffs, regulations, etc.—tend to be diffused over the entire population of taxpayers and consumers, any action costs each individual so little that it is not worth his time to organize in opposition. With the combination of concentrated benefits and diffused costs, well-informed and articulate interest groups (which contribute to campaigns) will dominate the political process and receive political favors.

3. *Shortsighted effects.* Politicians who must face the electorate every few years tend to be more concerned with the short run than with the long run. They will have little interest in policies that are efficient but that take time to produce results.

4. *Little incentive for candidates to account for individual preferences.* In the marketplace, consumers are generally able to tailor their purchases very closely to their own preferences and each individual gets the particular kind of product he wants. In the political marketplace, however, voters must decide on alternative bundles of governmental expenditure and tax proposals offered by competing politicians. There is no opportunity for individuals to pick some of one candidate's positions and some of another's and at the same time reject both candidates' positions on other issues. To capture as many votes as possible, the candidates' policy bundles tend to reflect a majority coalition, not the wishes of individual voters.

Drinking at the Political Water Trough

The potential for political processes to redistribute wealth from one individual or group to another opens the door for what economists call "rent seeking."[4] The rents referred to are not the payments made to landlords at the beginning of the month, but rather are returns to an asset that exceed the costs of producing the asset. For example, a professional baseball player like Alex Rodriguez earns a very large salary. Part of that is a return on investments he has made in perfecting and maintaining his skills. A much larger part, however, arises from the fact that people are willing to pay a great deal more to watch him play than they are willing to pay to watch others. This extra amount is a rent to the very special talents that Alex has. Such returns motivate entrepreneurs to always be on the lookout for scarce resources for which consumers will pay extra because the resource is so unique.

Rents can also arise because there are limits on competition such that profits are not competed away to the benefit of consumers and because resources are sometimes available for use by one party but paid for by another. Both of these too often characterize outcomes in the political process. Consider a politically allocated licence to provide cable television or internet service to homes in a particular area. If there are no alternative suppliers—satellite or antenna, for example—and if cable licence holders get their licences for free, they will receive rents because of the limited supply. Politicians or bureaucrats who decide who gets the licenses simultaneously determine who gets the rents. Subsidies do the same by giving the recipient of the subsidy "money for nothing" (McChesney 1997). Corn ethanol subsidies are a classic example.

Contrast the two types of rents in the context of water. Suppose that water from a river is being diverted to irrigate corn and that an acre-foot of water applied in this way results in a crop valued at $20. Now suppose that an entrepreneur sees a way of applying the same water to another crop such as grapes where its value is $50. By reallocating water from corn to grape production, the entrepreneur earns $30 in economic rents—returns above what the water is worth in its next most valuable use. Through this process, rents are created as the size of the aggregate economic pie expands.

Alternatively, suppose that the government is willing to tax its citizens to provide a subsidy for an irrigation project that produces grapes for which water is worth $50 per acre-foot. If water is free or costs less than $50 to the grape farmer, she earns rents equal to the difference. To garner such rents from the political process, grape farmers, much like ethanol producers, subdivision developers, environmentalists, or any other potential recipients of the largess, will spend resources trying to affect the outcome of political allocation. Those who win the competition for subsidized water capture rents at taxpayer expense, and those who lose will spend "money for nothing" (see McChesney 1997).

Entrepreneurs are always on the lookout for either type of rents. There is, however, a difference between seeking the two types of rents. In the marketplace, an entrepreneur who discovers a new resource—baseball player, diamond mine, or scenic mountain—and markets it is creating value for people willing to pay. The baseball fan, bridegroom, or homeowner with a view is paying to ensure that the asset is created or discovered and maintained.

Rent seeking through the political process comes with two potential adverse effects. First, there is no guarantee that the rents captured are worth the cost. Of course, the recipient will argue that they are, but the public choice problems enunciated above make it difficult to know whether there are real efficiency gains. Take, for example, the U.S. Bureau of Reclamation's Central Utah Project which delivers water over the Wasatch Range to the Salt Lake region. The annual cost to taxpayers to deliver the water is roughly $161 per acre-foot (U.S. Department of the Interior 2009), yet farmers generally pay less than $30 per acre-foot. With high crop prices, the rents accruing to irrigators can be substantial but unless the crops produced with an additional acre-foot of water are worth more than $161, a very unlikely prospect for something like alfalfa, the size of the overall economic pie shrinks.

A second consequence of political rent seeking is that the time and money put into rent seeking could be put to producing other things instead of redistributing wealth. In the end, wealth will be redistributed but nothing more will be produced. Of course, there are legal barriers to governmental redistribution such as the takings clause of the Constitution (Anderson and Hill 1980) and political barriers resulting from democratic elections, but it is impossible to plug all the holes in the rent-seeking dam, especially in the case of water where well-organized special interest groups have found ways to create and capture rents. If the rents are continually up-for-grabs through political reallocation, rent seeking will be even worse.

Political entrepreneurs seeking rents provide the demand for rent seeking, and politicians respond as suppliers meeting those demands in the name of the public interest. The rents arise because politicians and bureaucrats have access to the government treasury which is just as susceptible to the tragedy of the commons as any open access resource. They ask: what are the gains to my constituents and to me personally and what are the costs of capturing another dollar from the treasury? Discretionary control of money and resources fills campaign coffers and increases agency budgets with no guarantee that a dollar spent on one project will exceed the opportunity cost of spending it on another. By pursuing programs that concentrate benefits on the agency and its constituents while diffusing the costs over the general population, it will be possible to maximize support and minimize opposition. This is the essence of rent seeking.

Conclusion

The lesson from economics that good decisions require weighing of marginal costs and marginal benefits is critical to good water allocation. It does not follow from this lesson, however, that good decisions will flow from every and all institutional settings. Focusing totally on cost–benefit analysis in the political sector might make it appear that resource allocation decisions always accord with efficiency standards. This appearance, however, fails to account for how the information about values is generated and how it is used by decisionmakers (see Hayek 1945). After all, water in this country leaks from failing supply infrastructure, flows uphill to irrigate low value crops and quench the thirst of desert cities, and shoots out of decorative fountains to evaporate in the desert sun (Glennon 2009). These might very well be efficient allocations of water, but they are equally likely to reflect the inability of water bureaucracies to gather and process information on marginal costs and benefits and to supply water rents to well-organized constituencies, as public choice theory predicts.

In the chapters that follow, we will explore the incentives faced by market actors when water rights are relatively secure and tradable and those faced by politicians, bureaucrats, and their constituents when political institutions govern water allocation. For water markets to work, authority must be linked to responsibility through private property rights in a way that forces owners to take account of opportunity costs. If such rights are lacking, markets will fail to allocate water efficiently, begging the question: Will political allocation improve the situation?

Subsequent chapters will detail the evolution of water institutions, mainly in the United States, where the evolution can be characterized as beginning with markets underpinned by relatively secure water rights, transiting toward political allocation, and gradually shifting toward water markets. For now, we close with some data in Table 2.1 suggesting we are beginning to tap water markets in the American West. Between 1987 and 2007, more than 150 million acre-feet of water was traded among water users, with total expenditures reaching nearly $9 billion.

In one of the definitive explanations of the potential for water markets, Hirshleifer et al. (1960, 361–62) presaged this result, saying,

> Other things being equal, we prefer local to state authority, state to federal— and private decision-making (the extreme of decentralization) to any of these. Our fundamental reason for this preference is the belief that the cause of human liberty is best served by a minimum of government compulsion and that, if compulsion is necessary, local and decentralized authority is more acceptable than dictation from a remote centralized source of power. This is an "extra market value" for which we at least would be willing to make some sacrifices in terms of loss of economic efficiency . . . Even on grounds of efficiency, however, we have some faith that, the more nearly the costs and benefits of water projects are brought home to those who make decisions, the more correct those decisions are likely to be a consideration which argues for decentralization in practice.

TABLE 2.1 Market activity, 1987–2007

	Committed acre-feet	Total expenditures ($)
Arizona	22,668,859	837,843,211.20
California	41,827,751	4,789,242,789.00
Colorado	16,747,143	855,741,323.59
Idaho	10,130,111	108,970,002.90
Montana	599,336	5,518,829.70
New Mexico	2,711,755	93,549,513.82
Nevada	4,784,739	784,677,715.20
Oregon	9,727,495	219,994,636.40
Texas	38,373,901	907,882,623.20
Utah	5,028,801	139,232,808.80
Washington	3,970,445	24,149,687.45
Wyoming	711,654	5,715,517.95
Total	157,281,988	8,772,518,658.41

Source: Water Strategist, various years.[1]

Note: [1] Data were compiled from the *Water Strategist*, a monthly trade journal that tracks water transactions throughout the West. Although comprehensive, the dataset is not complete as not all transfers are reported or disclosed by the trading parties.

3

WHO OWNS THE WATER?

To understand the contemporary institutions that govern water allocation, we must begin with an understanding of how and why those institutions were adopted in the first place and how they have changed over time. Historically, water institutions evolved in response to the relative scarcity of local water resources. For example, when English settlers arrived on the Atlantic seaboard, water was abundant so there was little reason to worry about who owned the water or how it was used. Accordingly, the riparian doctrine's "reasonable use" standard is relatively unrestrictive. But conditions on the arid Great Plains and western frontier were much drier, making it necessary to hammer out new water institutions that would enable users to secure access to scarce water. The resulting divergence of water institutions in the wet East and dry West demonstrates how water scarcity drives institutional evolution and, in particular, the extent to which people define and defend natural resource claims.

People devote more effort to defining and enforcing property rights as resources become more scarce and therefore more valuable (Demsetz 1967; Anderson and Hill 1975; 2004). Take, for example, commercial fishing over the past few decades. As fish stocks declined throughout the world, fishermen made capital investments in bigger boats and more efficient gear. Their investments allowed them to claim the resource by catching it, but it also led to the collapse of several fisheries in a classic example of the tragedy of the commons (Gordon 1954). By assigning portions of the total allowable catch to individual claimants, individual fishing quotas emerged as a more economically and ecologically sound method of managing scarce fish stocks. But the fishing quota approach, like all property rights institutions, requires financial and political investment such that only a fraction of the world's fisheries have adopted them (Leal 2004).

Other factors contribute to increased definition and enforcement efforts. For example, if theft becomes more likely, property owners respond by installing more and bigger locks, installing burglar alarms, buying watchdogs, or hiring private protection agencies. Changes in the technology whereby property rights are defined and enforced can also change property institutions. An example is the introduction of barbed wire in the 1870s (Anderson and Hill 2004). To the open range cattle ranchers, barbed wire helped define boundaries, control grazing, rotate cattle on pastures, selectively breed livestock, and prevented rustling. Fence posts painted blaze orange in Montana provide a more modern example of efforts by landowners to define and defend their property rights. Increased hunting and fishing pressure on private land induced Montana farmers and ranchers to seek legislation allowing them to post land against trespassing by simply marking access points with orange paint (Anderson and Hill 1995).

Building water institutions that define and enforce water rights and allow their exchange is no different, albeit at times more challenging. When water becomes more scarce, it becomes more valuable, and it behooves people to invest in establishing property rights. But until the point at which the benefits exceed the costs of defining private water rights, the resource will be unowned or owned in common. The following examination of domestic water institutions explains how water's relative scarcity has shaped the cost–benefit calculus of defining private water rights. It also illuminates how well-defined, enforced, and transferable property rights can help resolve conflicts as water becomes more scarce.

Different Institutions for Different Climates

When people migrate to new places, it is common for them to take along their customs, culture, and legal traditions. In some instances, imported rules for allocating natural resources are well suited to resource endowments of the new location. So it was when European settlers applied the riparian doctrine to the abundant water resources of the eastern United States; the imported doctrine worked well. However, when the resource endowments of a new location differ from those of the home land, the imported customs, culture, and laws must be modified or completely abandoned to prevent resource waste and avoid shortages or crises. Such was the case when Easterners settled in the West. They found riparianism inoperable and soon replaced it with the prior appropriation doctrine, an institution premised on water scarcity.

The Riparian Doctrine of the East

Colonists coming from England to the New World brought with them English common law, including the riparian doctrine. The roots of riparianism lie in the common law tenet that waters are *publici juris* and that use is granted to those with access to the water, generally through landownership (Teclaff 1985, 26). Indeed, the origin of the word riparian comes from the Latin word *ripa*, meaning bank of a stream, and early courts considered riparian property to extend geographically to the rim of the land draining into the water body.

The earliest riparian laws were based on what is generally called the natural flow theory. This standard held that every landowner whose property was adjacent, or riparian, to a water body was entitled to have the water flow past his land "undiminished in quantity and unaltered in quality" (Caponera 2007, 125). Under the natural flow standard, riparian water users could divert and return water to the stream as it came natural to their land but could not appreciably change the water's volume, temperature, flow, or quality.

The natural flow standard was appropriate in areas where water was abundant and diverting water away from the water course was rarely necessary. Where precipitation levels were adequate to maintain crops and stream flows satisfied domestic, livestock, and power generation needs, maintaining the "natural flow" of a water course presented little difficulty. Thus, the riparian system first adopted in the eighteenth century was appropriate given the abundant water resources and few competing demands. At that time, water shortages and disputes over water use were "rare and short-lived," reflecting the resource's relative abundance (Dellapenna 2002, 1567).

As water demands increased, however, the natural flow theory became less practical. According to Caponera (2007, 126), "[e]ven in humid [states], conflicts developed as emerging industries, municipalities and agriculture began diverting water." Where and when such diversions altered the quantity or quality of water flow, the natural flow theory of the riparian doctrine stood as a barrier to economic growth. For a developing country like the nineteenth-century United States, such barriers were short-lived and the reasonable use standard was born.

The reasonable use doctrine developed as a necessary modification to the natural flow theory and now forms the foundation of most modern-day riparian doctrines. The reasonable use rule holds that a landowner adjacent to a water body, including streams, lakes, ponds, or marshes, is entitled to use of that water so long as the use is considered reasonable and non-injurious to other users. This correlative use right is not defined in terms of quantity or flow (as is the case under the prior appropriation doctrine), but depends upon a host of considerations including: the characteristics of the water body; the number of water users; the object, necessity and duration of the use; as well as the nature and extent of potential injury to other water users (Caponera 2007, 126).

The precise definition of reasonable use has changed over time and varies across states, but generally includes most agricultural, domestic, manufacturing, and recreational uses. Certain uses such as domestic use may be given preference as "natural wants" without regard to injury to other riparians. Generally, however, the reasonableness of any water use is determined by the harm caused to other users. According to the Minnesota Supreme Court,

> Each use is required to be beneficial, suitable to the watercourse and its economic and social value. If these requirements are met, reasonableness may require each riparian to put up with a minor inconvenience and to adjust to quantity of water used. [If a conflict occurs, a solution involves

consideration] of whether the first user's investment and other values are entitled to protection and whether the new user ought to compensate the former user for the loss of that which the latter gained. In most of the cases in which the plaintiff has suffered substantial harm through his water supply for a reasonable use being taken, the decision has been that the taking is unreasonable.[1]

This description accurately portrays the flexibility of the reasonable use standard and the uncertainty created by the resulting correlative right structure. Court or agency adjudication provides the only mechanism for prioritizing conflicting water uses, and that mechanism depends on a host of subjective evaluations. Consequently, security of right is limited or non-existent in most riparian states.

Related to this uncertainty is the fact that riparian landowners do not take ownership in water adjacent to their land; rather, they hold a usufructory right to a reasonable amount and to be free from unreasonable uses by other riparians. The state, generally through its constitution, possesses ownership over water and holds the water resources in trust for the people of the state. And though few riparian states require landowners make use of the water in order to preserve their right, the number of riparian states with forfeiture laws is increasing.[2]

The riparian doctrine is an institution premised on water abundance. Given that abundance, there was no foreseeable need for the early settlers to invest in a system that would establish and quantify ownership of water. Instead, they rationally decided water was to be shared coequally with other riparian landowners.

The shift from natural flow to reasonable use as the standard by which water use was governed, however, marked the first major institutional transition in response to water scarcity in the East. Another such transition is underway in many Eastern states as added competition for increasingly scarce water has again put pressure on balancing supplies and quantities demanded.

The Prior Appropriation Doctrine of the West

To the frontiersmen entering the Great Plains, it was clear that access to water would be a determining factor in locating a suitable farm or ranch site. Hence initial settlement patterns can be traced to river and stream bottoms (Anderson and Leal 2001, 33). If an individual found that a stream location was already taken, he simply moved to another water supply. Under these circumstances, the right to use the water accrued to whoever owned the stream bank and had access to it by virtue of position (Webb 1931, 433, 447).

It is not difficult to understand why riparian water rights, whether implicit or explicit, were applied first by the early frontiersmen. Early judges and lawyers in the West were only familiar with Eastern law and were inclined to transfer it to the West's legal institutions (Webb 1931, 447). In addition, land with adjacent water was abundant relative to the number of settlers; that is, water was not that scarce. As long as these conditions held, riparian rights sufficed. In other words, the benefits of changing from the riparian system were not sufficient to attract time and effort for establishing something new.

As scarcity increased, however, two factors affected the benefits and costs of altering property rights to water. First, mining technology required that water be taken from the stream and moved to non-riparian locations. Second, growing amounts of non-riparian agricultural land could only be made productive if irrigation water could be delivered to it. Riparian rules simply did not allow the necessary diversions for mining and irrigation.

Because the California mining camps were the first to feel pressure to divert water away from streams, it is not surprising that miners played an important role in the evolution of the prior appropriation doctrine.[3] The miners soon realized that gold was not only found along streambeds, where only a pan and shovel were needed to extract the precious mineral. When deposits were discovered several miles from water, it made economic sense to appropriate water from the streams. It "universally became one of the mining customs that the right to divert and use a specified quantity of water could be acquired by prior appropriation" (McCurdy 1976, 254). These customs had "one principle embodied in them all, and on which rests the 'Arid Region Doctrine' of the ownership and use of water, and that was the recognition of discovery, followed by prior appropriation, as the inception of the possessor's title, and development by working the claim as the condition of its retention" (Kinney 1912, 598). These customs soon evolved into more formal codes among the miners.

> Following a tradition of collective action on the mining frontiers of other continents, the miners formed districts, embracing from one to several of the existing "camps" or "diggings" and promulgated regulations for marking and recording claims. The miners universally adopted the priority principle, which simply recognized the superior claims of the first-arrival. . . . The miners' codes defined the maximum size of claims, set limits on the number of claims a single individual might work, and established regulations designating certain actions—long absence, long diligence, and the like—as equivalent to the forfeiture of rights. A similar body of district rules regulated the use of water flowing on the public domain.
>
> (McCurdy 1976, 236–37)

While there is no question that the original mining law was aimed at establishing private rights to water through appropriation, disputes over rights led to court cases which in turn led to conflicts with the riparian doctrine of common law. Bound by precedent and practicality, judges were torn between their training, which taught them that decisions ought to "conform, as nearly as possible, to the analogies of the common law," and the western tradition, which held that law "ought to be based on the wants of the community and the peculiar conditions of things."[4] The tensions between the riparian and prior appropriation doctrines are reflected in the finding by some courts that appropriative principles were "impractical" and the finding by others that cases "must be decided by the fact of priority." The result was an interesting and eventually harmful mix of eastern and western law. Webb captured the nature of the mix:

> The Easterner, with his background of forest and farm, could not always understand the man of the cattle kingdom. One went on foot, the other went on horseback; one carried his law in books, the other carried it strapped round his waist. One represented tradition, the other represented innovation; one responded to convention, the other responded to necessity and evolved his own conventions. Yet the man of the timber and the town made the law for the man of the plain; the plainsman, finding this law unsuited to his needs broke it and was called lawless.
>
> (Webb 1931, 206)

Cases were often complicated, but Judge Stephen J. Field of the California Supreme Court contended that "the courts do not . . . refuse the consideration of subjects, because of the complicated and embarrassing character of the questions to which they give rise."[5] The Field court continually worked to define and enforce rights in order to promote efficient markets. Even pollution, which frequently occurred in the mining process, was handled by having polluters pay damages to those users who received lower quality water. The impact of the Field court decisions are summarized by McCurdy as follows:

> By converting the possessory claims of so many trespassers into judicially-cognizable property rights, the California court effectively brought federal land-use policy into the realm of private, and, in some instances, constitutional law . . . Field believed that only the courts were capable of resolving allocation problems so as to simultaneously protect property rights, release entrepreneurial energies, and provide all men with an equal opportunity to share the material fruits of a vigorously-expanding capitalist society.
>
> (McCurdy 1976, 264–66)

The appropriation doctrine established ownership rights that were clearly defined, enforced, and transferable. Rights were absolute and not co-equal; first in time was first in right. As a result, markets were left to determine the value and allocation of water. The California courts asserted that "a comparison of the value of conflicting rights would be a novel mode of determining their legal superiority."[6] As McCurdy (1976, 257–58) stated, "Anyone might take and use water flowing on the public domain for any beneficial use subject only to the rights of any prior appropriators." The doctrine of appropriations gave no preference to riparian landowners, allowing all users an opportunity to compete for water and to develop far from streams. Appropriations were limited according to the means used for appropriation and the purpose of the appropriation.

The law that evolved in the West reflected the greater relative scarcity of water in the region. As the settlers devoted more efforts to defining and enforcing property rights, a system of water law evolved, which (1) granted to the first appropriator an exclusive right to the water and granted water rights to later appropriators on the condition that prior rights were first met; (2) permitted the diversion of water from the stream so that it could be used on non-riparian lands; (3) forced the appropriator of water to forfeit his right if the water was not used; and (4) allowed for the transfer and exchange of rights in water between individuals.

Organizing to Deliver Water

Under the prior appropriation institution, several types of private organizations developed to store and deliver water to where it was most productive. Though not every approach worked, much of the West's water development occurred because private companies invested in water development and delivery infrastructure. A brief history of these organizations and their performance sheds light on how water markets promoted water-use efficiency and minimized the "fighting" that typifies competition for water today.

Commercial Irrigation Companies

Many different forms of business organization were used to develop western irrigation. For the smaller projects, especially in the mining districts, individuals and partners had sufficient funds to undertake the necessary investments. The cooperative ventures of the Greeley, Anaheim and similar colonies irrigated thousands of acres by pooling resources to purchase land and construct irrigation systems.

In addition, corporations were formed to construct irrigation systems and sell the water to farmers. Initially, investors from eastern cities and Europe furnished capital in the hope of realizing huge returns once the settlers arrived and began farming. Diversion dams and canals capable of irrigating thousands of acres were built at great expense by companies. For example, the Northern Colorado Irrigation Company built the 85-mile High Line Canal to divert the Platte River; the Wyoming Development Company built a 2,105-foot dam and 2,380-foot tunnel to carry water from the Laramie River to irrigate more than 50,000 acres; the Pecos Irrigation and Investment Company built two dams and two canals, 30 and 50 miles long respectively, in southeastern New Mexico; the Idaho Mining and Irrigation Company constructed a diversion dam and canal to irrigate 600,000 acres; and the San Joaquin and King's River Canal and Irrigation Company built a canal capable of irrigating 16,000 acres (Dunbar 1983, 24; Worster 1985, 102).

Further expansion of commercial irrigation projects was thwarted by barriers to landownership. Most commercial irrigation enterprises were built on watercourses flowing through federal lands. However, the riparian public lands were not generally available for ownership to secure private investments. Some investors obtained grants of land from the federal and state governments, but most of the grants were opposed and defeated by those who feared monopoly by such enterprises (Pisani 1992, 91–92). When the corporations began to face financial difficulties, the optimism of their investors disappeared and no further funds for irrigation development could be found (Huffman 1953, 82).

Eventually, most of the commercial irrigation corporations went out of business for several reasons. First, their developers were ignorant of the actual amounts of water carried by streams and the amount of water necessary to irrigate a given quantity of land, as well as of the effect of the often harsh and inconsistent climate on the water supply (Baker and Conkling 1930, 9–10). Often, those first to arrive on the frontier had an unbridled optimism. In the race to establish a claim on the frontier, they arrived far ahead of a time when the land and water could really generate a profit.[7] Second, the corporations underestimated the cost of constructing and maintaining irrigation works (Alston 1978, 109). When settlers did not arrive in droves and immediately begin buying water for farming as expected, overhead costs quickly overwhelmed the irrigation enterprises. In the absence of further investment capital, they were driven into bankruptcy. Moreover, states began passing laws allowing county commissions to regulate the rates commercial companies could charge for water, which prevented them from recapturing their costs. Good came of the failure of these enterprises, however, in that thousands of acres were irrigated by the systems the companies built and many of the irrigation works were taken over by settlers who formed mutual companies or irrigation districts and continued their operation (Dunbar 1983, 24–27; Pisani 1992, 106).

Development corporations fared better than the large commercial irrigation companies because they internalized the costs and benefits. They purchased large parcels of land, built irrigation works to provide water for farming, formed mutual irrigation companies, and then subdivided and sold the parcels along with a proportionate number of shares in the mutual company. By the time all the land was sold, the purchasers owned the water company. Development corporations were so successful that thousands of acres in California were irrigated in this manner (Worster 1985, 100–102).[8]

Mutual Irrigation Companies and Cooperatives

Following the decline of the commercial irrigation companies, a more localized use of the corporate structure succeeded in amassing and structuring capital for the construction and operation of Western irrigation systems (Bretsen and Hill 2007). Mutual irrigation companies, or "mutuals," were local corporations formed by local irrigators for the purpose of providing water at cost to the shareholders rather than profits for distant investors (Dunbar 1983, 24–26). Mutuals captured the benefits of scale economies in irrigation with lower transaction costs than were borne by commercial irrigation companies and without the post-contractual opportunism created by separating the water suppliers from the water users (Bretsen and Hill 2007).

The structure of mutuals was especially significant since it further contributed to the operation of private markets in the West. Shares of stock were issued to each member representing a percentage of the water supply, which was prorated each year according to stock ownership. The shares could usually be rented or sold and transferred anywhere within the service area of the mutual. The price of the shares was determined by the marketplace. Members either held on to their water rights or deeded them to the mutual (Smith 1984, 167; Pisani 1992, 101).

Mutual companies either built their own irrigation works or purchased those of other irrigation companies.[9] They levied assessments against shares of stock for construction, maintenance and operation, and borrowed money secured by liens on the members' land for larger projects (Baker and Conkling 1930, 396–97). By using member assets as collateral, mutuals could enter capital markets to obtain the investment funds necessary to develop irrigation projects. The transferability of stocks ensured that water could be moved to higher valued alternatives, further ensuring the success of the operation. These features, combined with the security of rights provided by the prior appropriation doctrine, stimulated an effective marketplace.

The efficacy of mutuals as a way of organizing water rights and investment is evidenced by the fact that mutuals became the dominant institution for irrigation management throughout the West and especially in California, Colorado, and Utah. Though a few were large, such as the Twin Falls Canal Company of Idaho with 204,000 acres and the Salt River Valley Water Users Association with 228,000 acres, most were small, serving neighborhoods or communities. As of 1925, the average size of most mutuals was between 1,000 and 2,000 acres (Baker and Conkling 1930, 396; Huffman 1953, 72).

Given the challenges faced by the early settlers to the West to develop water supplies, the contributions of private reclamation projects should not be underestimated. By 1910, over 13 million acres of land in the West were irrigated by private ventures. Between 1900 and 1910, the number of irrigated acres grew by 86.4 percent, with private enterprise accounting for almost all of the increase (see Table 3.1) (Golze 1961, 6).

TABLE 3.1 Private irrigation development in 17 western states, in acres

Census	Total irrigated acreage	Furnished government water	Private development
1890	3,631,381	–	3,631,381
1900	7,527,690	–	7,527,690
1902	8,875,090	–	8,875,090
1910	14,025,332	568,558	13,456,774
1920	18,592,888	2,388,199	16,204,769
1930	18,944,856	3,049,970	15,894,886
1940	20,395,043	3,800,239	16,594,804
1950	24,869,000	5,700,000	19,169,000

Source: Golze 1961, 14.

While individual and partnership efforts were responsible for irrigating the bottom lands along rivers and streams by means of crude diversion dams and ditches, it was cooperatives and other business organizations that developed the fertile benchlands and non-riparian ground (Dunbar 1983, 20). By 1920, cooperative ventures, including incorporated and unincorporated mutual companies, and irrigation districts had surpassed individuals and partnerships in number of acres irrigated (see Table 3.2). Even though public development greatly increased after passage of the Reclamation Act of 1902, private development continued to provide a significant portion of new irrigation development.

In spite of the financial failure of many of the initial private attempts at irrigation development, Elwood Mead, one of the leading irrigation experts of his day, wrote in 1903 that corporate capital investment in canals

> has been the leading factor in promoting agricultural growth of the western two-fifths of the United States. It has been the agency through which millions of dollars have been raised and expended, thousands of miles of canals constructed, and hundreds of thousands of acres of land reclaimed. It has been the chief agency in replacing temporary wooden structures by massive head-works of steel and masonry, and, by the employment of the best engineering talent and the introduction of better methods of construction, has promoted the economy and success with which water is now distributed and used.
>
> (Mead 1903, 57)

In discussing the contractual benefits of the Colorado ditch companies, Mead concluded that "if the water of streams is public property, the public should show

TABLE 3.2 Area irrigated in 17 western states, by type of enterprise

Item and type of enterprise	Primary enterprises (acres)		
	1920	1930	1940
Individual and partnership	6,448,663	6,038,835	6,906,738
Cooperative, incorporated	6,569,690	6,271,334	5,706,606
Cooperative, unincorporated	–	–	907,242
Irrigation district	1,822,887	3,452,275	3,514,702
Reclamation district	–	–	59,052
Commercial	1,635,027	999,838	855,166
Bureau of Reclamation	1,254,569	1,485,028	1,824,004
Bureau of Indian Affairs	284,551	331,840	515,765
State	5,620	11,472	16,995
City and/or sewage	40,146	121,218	83,457
Other	531,735	233,016	5,316
Total	18,592,888	18,944,856	20,395,043

Source: Golze 1961, 99.

the same business ability in disposing of its property as those to whom its control is transferred. Colorado can learn something about the management of the water of streams by studying how canal companies dispose of the water which they appropriate" (Mead 1903, 167).

Irrigation Districts

In the late nineteenth century, water development and management in the West began to shift from predominately private to the public sector. California's Wright Act, passed in 1887, was the first legislation creating irrigation districts and granting them limited governmental powers as a means to finance and manage irrigation systems[10] (Huffman 1953, 71). By the early twentieth century, irrigation districts had emerged as the leading organizational structure for water storage and delivery in the West. The majority of these new districts had been reorganized from mutual or commercial irrigation companies (Hutchins et al. 1953).

As quasi-municipal corporations and political subdivisions of the state, the irrigation districts had powers that went beyond those of private corporations. Districts could issue bonds and secure them by attaching liens to all lands within the district. And revenue to pay interest on the bonds and for maintenance and operation of irrigation works was raised from tax assessments levied against that same property (Bretsen and Hill 2007). All property within the district was included in the district, regardless of whether the owner wanted to participate and whether the property was irrigable.[11] This allowed irrigation districts to overcome the free-rider and hold-out problems that had plagued mutual irrigation companies. Moreover, districts had the power to condemn water rights to provide a unified distribution system (Chandler 1913, 133; Baker and Conkling 1930, 395; Huffman 1953, 73–75).

Though they were established as public institutions, irrigation districts were formed to benefit a specific group of private landowners, meaning irrigation districts could "affect public colors when advantageous, but resort to private camouflage when needed" (Tarlock et al. 2002, 775). This provided irrigation districts a competitive advantage over private irrigation and helped facilitate the takeover of private enterprises (Bretsen and Hill 2007). Combined with the exclusive ability to receive Bureau of Reclamation water (Leshy 1982, 359–60), it is little wonder irrigation districts outperformed the purely private irrigation institutions (Dunbar 1983, 33–34; Pisani 1992, 103–104).

Regulated Riparianism

The stark contrast between the riparian doctrine of the East and the prior appropriation doctrine of the West demonstrates the role of scarcity in the evolution of water institutions. That scarcity-driven evolution is continuing today as population growth and prolonged drought are undermining the notion of water abundance in the East and forcing several Eastern states to reconsider traditional riparian principles. Evidenced by widespread water shortages and increased conflict over uncertain supplies, the riparian doctrine has proven to be poorly equipped to allocate water efficiently to meet new and growing demands.

Reversing Horrace Greeley's admonition to "go West," principles from the rights-based prior appropriation system have migrated East. Rather than better defining and enforcing water rights to encourage water markets, however, Eastern states have modified the traditional riparian rights system into "regulated riparianism." In 1997, the Water Law committee of the American Society of Civil Engineers developed a comprehensive guide (updated in 2004) to assist state governments in transitioning to a regulated riparian system of water allocation (Dellapenna 1997). The Regulated Riparian Model Water Code is a comprehensive body of proposed legislation addressing many aspects of water allocation, including water quality, conservation and minimum flows, conflict resolution, drought management strategies, and transboundary water issues. Although states vary in the degree to which they have integrated the suggested statutory framework, it provides a generalization for discussing regulated riparianism.

In contrast to the traditional riparian or the appropriative rights systems, the regulated riparianism model treats water as "a species of public property rather than as either common or private property" (Dellapenna 2002, 10). Under this altered system, water resources are no longer allocated and shared coequally among riparians. Instead, water allocation and use decisions are made by the state. States believe centralized control of water will improve productive and efficient water allocation, provide a means of reallocating water in times of shortage, and improve overall, the protection of the public interest including water for environmental and recreational uses.

One important characteristic of regulated riparianism is how it allocates water. Under the traditional riparian rights system, water rights are established through appurtenancy to water courses and disputes are settled through common law judicial processes. In contrast, under the regulatory framework, water is allocated through a state-administered permit process and disputes are preemptively addressed through statutory requirements that are designed to eliminate the risk of future conflicts.

To be clear, not all water uses require a permit. Although states vary in their requirements, generally only large consumptive water users are required to go through the state permitting process. Florida, for example, maintains comprehensive permitting statutes requiring state approval for most consumptive, including agricultural, municipal, and industrial applications, as well as for large (in excess of 100,000 gal/day) groundwater withdrawals. Comparatively, in Massachusetts, only large, new consumptive uses require a permit.[12] Further, many states include provisions that exempt (or grandfather) users that were withdrawing water at the time the statue went into effect (see Dellapenna 1991). For example, in Virginia, permits are not required for uses in place before July 1, 1989.[13]

Unlike traditional riparian rights that continue ad infinitum, permits are generally issued for a specific term, typically 3–20 years (Dellapenna 2002), and are reviewed upon expiration. Temporal control over water resources provides agencies the ability to adjust water allocation in response to changes in supplies or uses. For permit holders, there is no guarantee of renewal as the state may determine the use is no longer "reasonable" or is harmful to other users, and the permit may be altered or denied. The adverse consequences of temporal uncertainty are discussed in Chapter 6.

The criterion for issuing permits is complex and varies by state and over time as water conditions fluctuate. State agencies interpret the statutes governing water allocation decisions based on current and projected water availabilities, preferred uses, environmental conditions, and how best to serve the public interest. In other words, water allocation decisions are left to the state agency which determines whether or not the proposed use is "reasonable." This can be particularly challenging and subjective considering "the factors to be considered include both abstract questions of the social utility or value of the proposed use and relational questions of the relative value of the proposed use compared to other existing or planned uses" (Dellapenna 2004, 6R-3-02). Further, the agency must "consider impacts on users dependent on other hydrologically interconnected water sources and on users in other water basins and in other States" (Dellapenna 2004, 6R-3-02).

The permitting system does share some similarities to the way in which water is allocated in western prior appropriative states. First, permits may be issued to non-riparian users, overriding the traditional appurtenancy requirement for water usage and providing for water developments away from natural water courses. The Model Code recognizes "there is no reason to believe that limiting the lawful use of water to riparian is necessarily always either efficient or ecologically sound, let alone that it will necessarily serve the public interest" (Dellapenna 2004, 6R-3-02). Second, if water supplies are insufficient to meet all demands, water allocation may follow use priorities where certain uses are given preference to others. For example, in much of the West and in eastern regulated riparian states, domestic water uses such as for human consumption or sanitation take precedent over agricultural or industrial uses in times of inadequate supplies. Third, regulated riparian water permits are quantified. Permit terms and conditions generally define at the least, the authorized use, amount of withdrawal, and timing of use. More detailed information including the amount of water consumptively used,[14] return flows, and special conditions regulating future uses may also be included, providing state agencies with a detailed inventory of water resources and where and how they are being used. Moreover, states may require users not subject to obtaining a permit to periodically register their use with the appropriate state agency (Dellapenna 2004, 6R-1-06). And lastly, permits may be transferable through market transactions. The Model Water Code encourages states to permit voluntary water right trades that are subject to the protection of other permit holders and the public interest (Dellapenna 2004, 1R-1-07), analogous to the tenet of "no injury" common in Western states. Generally, however, states have been reluctant to adopt the appropriate statutes or clarify how trades would occur (Dellapenna 2004). Moreover, as discussed in Chapter 6, the institutional components necessary for efficient trades are often missing in eastern states.

One key aspect of the permitting system is that it was designed to give states the ability to adjust water use and allocation over time, notably in the case of water shortages or water emergencies. State agencies are empowered through regulated riparian statutes with the legal authority to restrict uses. Paired with a centralized and detailed database of permitted water users and quantities, states may "restrict any water right by adjusting the terms or conditions of the permit" quantitatively or spatially to meet allocation goals—primarily to maximize the public interest as determined by the agency (Dellapenna 2004, 7R-3-02). In restricting permit conditions, the state agency must also consider previously established drought management strategies and comply with use preferences. Moreover, water restrictions may be placed on water users not required to have a permit, generally those relying on smaller water sources or small-scale users. The Model Code declares "[d]uring a water shortage or water emergency, the State Agency is empowered to restrict withdrawals for which no permit is required, or to allocate water to and among such uses to alleviate the water shortage or water emergency" (Dellapenna 2004, 7R-3-05).

Since the 1950s, Hawaii and 17 of the 31 states east of the 100th meridian have adopted some form of regulated riparianism (Dellapenna 2002, 1567). Many of the remaining eastern states seem to be moving toward greater state involvement in water allocation and regulatory management, especially as a means of allocating water during times of shortage (Miano and Crane 2003). Many scholars also believe centralized administration and regulation of water resources is a suitable path for eastern states given the challenges they face and, in general, is an improvement over the traditional common law system (see Dellapenna 1991; 2001; 2004; Marcus and Kiebzak 2008). Without exception, however, they acknowledge that regulated riparianism is neither a panacea to eastern water issues nor a system without costs. Chapter 6 provides a comprehensive look into the shortcomings of regulated riparianism and suggests how it may be used in the context of markets to improve water allocation.

Conclusion

Water institutions define how we allocate water and our capacity to adapt to scarcity and uncertainty. In recent decades, burgeoning populations, erratic precipitation, and new demands for water have been a true test of the efficacy of our current allocation systems. Moreover, these challenges have revealed institutional short-comings and motivated institutional changes, especially in historically humid regions where water resources are no longer abundant. To mitigate the risks of water shortages and water crises, water institutions must adopt an efficient and equitable means of allocating water among competing uses.

The history of the American West illustrates that water institutions can facilitate voluntary exchanges of water capable of moving it to higher valued use. During the last half of the nineteenth century, miners and farmers laid the foundation for water markets by creating the prior appropriation doctrine. Under this legal system, water rights were defined and enforced and made transferable. In part, it was a sense of justice that led the early settlers to allocate water rights on the basis of "first in time, first in right" (Lueck 2003). Reality also forced Westerners to adapt their water institutions to the aridity west of the 100th meridian. Given varying degrees of water scarcity, therefore, it is not surprising that the most arid states— Montana, Wyoming, Colorado, and New Mexico, where rainfall averages 15 inches per year—totally abrogated the common law riparian doctrine, while the less arid states—North Dakota, South Dakota, Nebraska, Kansas, Oklahoma and Texas— only modified it.

The prior appropriation system has evolved into a legal framework capable of encouraging individual actors in the marketplace to efficiently allocate water. Authority is linked to responsibility, giving water owners the incentive to seek out the highest and best uses of the resource. With an efficient set of water institutions in place, individuals undertook projects to deliver water where it was demanded.

Historically in the East, abundant water resources made adaptation of the prior appropriation system unnecessary, but quoting singer Bob Dylan, "the times they are a-changing." Water demands now often exceed supplies, precipitation patterns have shifted, and new demands for recreational and environmental uses of water have emerged. The overwhelming response to these changes has been to regulate control of water resources, under the assumption that state agencies are better equipped to make water allocation decisions for current as well as future water users. As we will discuss in Chapter 6, the regulatory approach is already straining to meet changing demands, suggesting that it might be worth considering reforms more in keeping with those hammered out in the "wild West."

4

WATER IS FOR FIGHTIN'

In May 2001, press reports estimated 13,000 people gathered in downtown Klamath Falls, Oregon, in protest of the federal government's decision to cut water supplies to most of the area's 1,400 farms. Instead of irrigation, the water was to be used to maintain lake levels and stream flows for endangered sucker fish and coho salmon. In a ceremonial act of defiance, area farmers, ranchers, and their families formed a mile-long "bucket brigade" to bring water, one bucket at a time, from the Upper Klamath Lake, down main street, and into the basin's principal irrigation canal. Others carried posters that read "Call 911, some sucker stole my water," "Feed the feds to the fish," and "Farmers: the endangered species." Later in the growing season, after much political wrangling, the Bureau of Reclamation authorized the release of some water to farmers. Unfortunately by then it was too late; the farmers lost an estimated $157 million in revenues from lack of water for irrigation (Hathaway 2001, 14).

The following year, despite continued drought conditions and strong opposition from environmentalists and Indian tribes, the Bureau of Reclamation restored water deliveries to area farmers, reducing the amount of water available for fish and reversing course from the previous year. Between mid- and late September, the Klamath experienced the largest fish kill ever recorded in the West when more than 33,000 salmonids (Chinook, coho, and steelhead trout) died within the lower 36 miles of the river. The primary cause of the fish kill was a pathogen (gill rot disease), though lower stream flows and warmer water temperatures may have played a role (California Department of Fish and Game 2004).

The Klamath Basin debacle was not due to the finite supply of water in the river per se, but to the allocation of overlapping and unclear legal rights to that water. The problem of too many claimants fighting for scarce water began in 1905, a century before the fish kill, when the Bureau of Reclamation asserted dominion over all unappropriated water in the Klamath and Lost rivers and their tributaries.

At that time, there were already private rights established through prior appropriation as well as tribal claims of unspecified amounts, so it is unclear just how much water the Bureau of Reclamation actually claimed when it began the Klamath Reclamation Project. The application of the Endangered Species Act cast additional uncertainty over the security and priority of vested water rights. This uncertainty precluded voluntary trades between the various claimants, leaving water allocation decisions to the political process where one party's gain is another's loss.

Unclear and overlapping legal claims have set off an even larger water war in the Sacramento–San Joaquin River Delta. In this inland estuary of northern California, the federally protected delta smelt (*Hypomesus transpacificus*) (U.S. Fish and Wildlife Service 1993, 12863) has stalled two of the world's largest water conveyance projects, the Bureau of Reclamation's Central Valley Project and California's State Water Project. The curtailment of delta deliveries reduces Central Valley agricultural revenues by an estimated $1.2 to $1.6 billion[1] annually (Howitt et al. 2009), an economic impact that supports the minnow-sized delta smelt's reputation as "the most powerful player in California water" (Boxall 2011).

At the core of this controversy is the same-use conflict that surfaced in the Klamath Basin: farmers and fish cannot use the same water at the same time. And like the Klamath, the use conflict escalated to a full-blown water war when competing user groups asserted overlapping legal claims to the resource. As David Slade (2008, 227) explains, the problem in the delta is a "classic clash between the doctrine of prior appropriation, and its 'first in time, first in right' water rights, and the Public Trust Doctrine's mandates for environmental protection of watercourses." Without a clear priority between these competing claims, neither side has an incentive or even a basis from which to negotiate. Consequently, protracted litigation and political maneuvering will ultimately determine the allocation of delta water.

As these examples illustrate, the current system of water rights in the West is characterized as much by conflict as cooperation. This chapter explains how litigation and lobbying became so prominent in western water allocation despite the institutional foundations for private water trading that formed on the frontier. This chapter begins with an analysis of the most common arguments against water markets. It then explains how restrictions on the market, called on to remedy these alleged failures, have done more to hinder than enhance the efficiency of water allocation. The chapter concludes with a discussion of how federal reclamation has crowded out private water marketing and further politicized water allocation.

Alleged Market Failures

Chapter 3 described how the doctrine of prior appropriation emerged from the mining camps and agricultural communities on the western frontier. This system of clearly defined, secure, and transferable water rights was well suited to the West's aridity because it allowed users to allocate water where it was most productive. During the latter half of the twentieth century, however, the basis of western water allocation shifted away from private contracting toward a complicated array of administrative procedures, legal doctrines, and government subsidies. The impetus for this shift to centralized water allocation was the alleged failure of water markets and, in particular, concerns over monopoly distortion, inadequate capital, and externalities. Evidence from the frontier suggests these claimed market failures were overstated, if they existed at all, and did not justify the restrictions on water marketing that persist today.

Monopoly Distortion

Early water reformers feared that private water suppliers would monopolize supply and charge high prices for the resource. Journalist and writer, William Ellsworth Smythe, for example, worried:

> [i]f we admit the theory that water flowing from the melting snows and gathered in lake and stream is a private commodity belonging to him who first appropriates it, regardless of the use for which he designs it, we have all the conditions for a hateful economic servitude. Next to bottling the air and sunshine no monopoly of natural resources would be fraught with more possibilities of abuse than the attempt to make merchandise of water in an arid land.
>
> (quoted in Alston 1978, 128)

Renowned western explorer John Wesley Powell recognized that the cheapest and most dependable source of water was from water companies selling at a profit, but he was also concerned with "the danger of an evil monopoly which would charge an exorbitant price and force the homesteaders to pay a heavy tribute" (quoted in Alston 1978, 129). This concern continues today, and it has contributed significantly to the government's administration of water resources.

The history of western irrigation, however, brings into question several of the assumptions upon which the fear of a water monopoly is based. Primary among these is the assumption that private water companies had the ability to exercise monopoly control over water users, to do so profitably, and for an appreciably long time. Though commercial irrigation companies of the frontier era displayed characteristics of a natural monopoly—their product was a necessity of life, the business had high economic and legal barriers to entry, and it was capital-intensive with high fixed costs with increasing returns to scale that led to declining unit costs—at least three factors substantially curtailed the market power and profitability of these companies (Alston 1978).

The first factor concerns the distinction between a simple and a bilateral monopoly. Even if a commercial irrigation company was the only water supplier to farmers in a particular river basin, those farmers were in turn the only viable customers of the irrigation company. Consequently, neither the water supplier nor the water user had a disproportionate amount of bargaining power over the transaction (Bain et al. 1963, 310). The price of water was often negotiated by both the suppliers and the users; private irrigation companies that attempted to exercise monopoly power by setting their prices substantially above marginal costs could expect to enter few, if any, contracts (Alston 1978).

The second factor limiting the ability of private irrigation companies to exercise monopoly power is closely related to the first: these companies had limited means of withholding their product from the market. In the event that an irrigation company set an exorbitant price that few irrigators would accept, the company would be forced to either hold the water behind a storage facility or release it downstream where courts held the water was free to be appropriated by others. Though companies may have been able to hold back enough water to keep prices up temporarily, the water would eventually have to be released. As a result, the monopolist strategy of increasing prices by restricting output was difficult to employ and rarely successful (Alston 1978).

The third factor which undermines the validity of water monopoly concerns in early western irrigation is perhaps the most persuasive, and that is market competition. Historical accounts describe a highly competitive market, one in which water users often could choose between water suppliers and influence the terms of the water supply contract.[2] In many cases companies competed with one another to provide irrigation water to the same region, if not the same acres (Dunbar 1983). Advances in groundwater irrigation technology gave irrigators yet another option, further reducing the prospect of a private water monopoly.

These limitations on the monopoly power of private water companies persist today, as does extensive regulatory oversight of water pricing (National Research Council 2002). Nonetheless, modern-day proponents of centralized water allocation insist—as did Powell and Smythe at the dawn of western irrigation—that a public water supply monopoly is preferable to privatization. This insistence is premised on the fear that privatization unequivocally results in higher water prices—a fear that has no empirical support—as well as the sentiment that water is a fundamental right ill-suited to market provision by profit-driven companies (Segerfeldt 2005). This aversion to privatization ignores the consequences of a public monopoly in water service provision that were described in Chapter 2: deferred maintenance, stagnant infrastructure expansion, and poor customer service.[3]

For monopoly distortion to constitute a legitimate water market failure, two conditions must be met. First, private water companies must have been capable of exercising monopoly power over water users and, second, water allocation by government agencies must have improved the situation by alleviating the monopoly distortion. The history of private irrigation companies in the West casts serious doubt on the first condition; and the poor performance of publicly owned and operated municipal water service similarly undermines the second.

Inadequate Capital

Another common argument made for government intervention in western water allocation was that capital markets were unable to provide the investment funds necessary for large irrigation and municipal water projects. Regarding frontier irrigation, Alfred Golze (1961, 12) explained "while private enterprise had managed to bring under successful irrigation an impressive and substantial acreage of land, a point had been reached where further development would need stronger support by the Federal and state governments." The turning point Golze describes was reclamation on such a scale as the Columbia Basin Project (Grand Coulee Dam) and the Boulder Canyon Project (Hoover Dam); and, indeed, private capital was not a sufficient funding source.

The insufficiency of private capital to fund every project pursued under the 1902 Reclamation Act does not indicate market failure so much as market success. These were bad investments as evidenced by the Bureau of Reclamation's ability to recover only a fraction of the capital costs and operation and maintenance costs for most federally funded reclamation projects (U.S. Congressional Budget Office 2006, 5–7). This poor financial performance is not surprising; many federal reclamation projects were sited in locations poorly suited to human habitation which, consequently, had previously been rejected by private investors. Las Vegas presents the most obvious example. The dirt where countless casinos, convention centers, and strip malls now sit was once a "whistle-stop on the Union Pacific Railroad," far from water sources (Glennon 2009). The Bureau of Reclamation changed that with its construction of Hoover Dam, a project that private investors were unwilling to fund and one that, by measure of the conflict surrounding Las Vegas' water consumption today, exacerbates water conflict in the region. In other words, the inability or unwillingness of private water companies to fund large-scale water development projects may reflect shortcomings of the projects themselves rather than shortcomings of the marketplace.

Externalities

Another alleged market failure used to justify centralized water allocation was the potential for water markets to generate harms external to the water trading parties. For instance, if an irrigator leased water to a mining operation that then used the water for pollution abatement, downstream users could suffer an economic harm from the diminished water quality. Similarly, if a farmer sold water to a landowner located further from the watercourse, the return flows to the watercourse could be delayed, reduced, or eliminated to the detriment of downstream water users. Water trading on the frontier did occasionally generate this sort of conflict (McCurdy 1976) and unless the property rights of downstream users specified otherwise, water traders were free to ignore the external harms.

Internalization of these harms required clarifying the definition of water rights and reducing the costs of their enforcement. As Nobel laureate Ronald Coase (1960) explained, if property rights are well defined and enforced and the costs of proving damage are relatively low, rights holders will enter agreements that promote efficient resource allocation. Although the water rights that emerged on the western frontier were not always defined with sufficient clarity to prevent downstream harms, courts were moving in that direction as use conflicts arose. Concerning water quality disputes, for instance, courts "issued injunctions when debris buried the claims of miners below, destroyed the growing crops of preemption claimants, filled irrigation ditches and poisoned their fruit trees, or split the hoses of hydraulic miners downstream" (McCurdy 1976, 262). In *Jennison v. Kirk* (1878), a California case in which a miner was held liable for damages when his debris washed away the ditch of another appropriator, Judge Stephen Field ruled that "no system of law with which we are acquainted tolerates the use of one's property in that way so as to destroy the property of another."[4] Such clarifications of water rights, combined with the common law doctrine of nuisance, gave water users a legal mechanism for internalizing externalities.

Despite the incremental success of this property rights approach, western states embraced an administrative solution as a check against harm to other property owners. Specifically, states began requiring administrative approval of all proposed water right transfers and, in particular, evidentiary proof that the proposed transfer would not harm other water rights holders. Hartman and Seastone (1970, 2–3) describe how concern over externalities has manifested itself in water institutions:

> [T]he problem of externalities is widespread, and various organizational arrangements and regulatory measures have been adopted or proposed to cope with it. Laws have been written and established by courts to protect the third parties in water transfers. Special districts have been formed to internalize some of the externalities. The general tendency in institutional development has been to modify market procedures or completely replace them.

As discussed below, substituting bureaucratic oversight for clearly defined and enforceable water rights helps a little to internalize the economic harms of water transactions, but goes a long way toward creating rent-seeking opportunities and inefficient allocations.

The prior appropriation system of water rights that evolved during the nineteenth and twentieth centuries was by no means perfect. When downstream users claimed return flows, conflicts resulted as upstream users changed the amount or location of diversion, leaving less water to flow downstream. Capital markets were in their infancy, so some economical water projects were probably not funded. With scale economies and imperfect capital markets, owners of private delivery systems no doubt had some market power. In general, frontiersmen were experimenting with new institutional arrangements that were imperfect but evolving to address the increasing demands on the West's water resources.

It is doubtful, however, that these imperfections in the system justify the extent of governmental intervention that occurred during the twentieth century. State allocation and administration replaced private ordering and trade. Overlapping claims established through the public trust doctrine, navigational servitude, and federal reclamation undermined the clarity and priority of appropriative rights. As a result, it was not long before the decentralized institutions that encouraged private contracting and efficient water allocation were replaced by cumbersome bureaucracy and constant litigation.

Divested Rights, Diluted Markets

In the late nineteenth century, the American West was evolving an effective system of water rights which created relatively well-defined and enforced rights, given the technology of the time, and that allowed trading to move water to higher valued uses. Ultimately however, rent-seeking reformers used the above allegations of market failure to divest private water rights and dilute water markets as the primary mechanism of water allocation.

California courts, under the direction of Judge Field, recognized the potential for "using the organized power of the community to divest the equitably acquired claims of men who had evinced a growth inducing 'incentive to improvement'" (McCurdy 1976, 265). In spite of the court's insight, the divestiture of water rights began in the late nineteenth century when restrictions were placed on the prior appropriation doctrine. State laws recognized prior appropriation rights, but most western state constitutions and statutes placed the physical water under the umbrella of public ownership. Appropriators under those systems received only a usufructuary right—a category of property rights that allows for the use of, but not actual ownership, of the resource. As a result, state legislatures could declare that the *corpus* of water belonged to the state and its general citizenry. This distinction opened the door for regulating water use on the grounds that the public has an interest in the resource which cannot be given in totality to private owners.

State Administration

The doctrine that had evolved through the decentralized actions of miners and irrigators slowly degenerated to the status of state-controlled permits and licenses. As early as 1929, Moses Lasky declared that the principle of appropriation had reached its zenith. The water rights that evolved in the quasi-anarchistic setting of the frontier were replaced by permits to use state-owned water, with decisions on water use ultimately made by state officials. Lasky (1929, 162) argued that these changes were causing a move away "from various forms of extreme individualism and vested property rights of substance in water to . . . the economic distribution of state-owned water by a state administrative machinery through state oriented conditional privileges of user."

Putting water allocation decisions in the hands of the state creates an opportunity to move water through political rather than economic channels, opening the door to rent seeking and inefficient allocation. With an agency or court empowered to allocate or reallocate water, users have an incentive to invest in influencing the agency's or court's decision. In the state of Montana, for example, the Department of Natural Resources and Conservation is empowered to reserve unclaimed water for future uses. Only governmental entities such as cities, counties, state and federal agencies, and irrigation districts may apply to reserve water for existing or future beneficial uses based on anticipated needs. Once reserved, local conservation districts authorize the use of the water and can reallocate reservations at a later date if they feel that the water is needed more for another use. Needless to say, the process for determining future needs involves considerable discretion on the part of the agency and conservation districts. Therefore, the competing governmental entities invest considerable time and money to influence the board's reservation decisions. Ultimately the rent-seeking game is zero-sum, as one party's gain is another's loss.

A similar problem arises when the state or courts can influence how water is reallocated by placing restrictions or limitations on the transfer of water rights. At a time when fewer and fewer sources of water are available for appropriation, meeting additional and changing water demands requires the movement of water away from traditional uses to now higher valued uses. Yet, restrictions on transfers remain widespread despite the clear economic, and even environmental benefits that could otherwise be realized in the absence of such restrictions.

In Wyoming, for example, the state prohibits private entities from acquiring water for environmental purposes such as instream flows for fish or wildlife. Rather, the state Game and Fish Department has sole discretion over the provision of instream flows through transfers, irrespective of whether or not a private individual values free flowing streams over traditional water uses. Hirshleifer et al. (1960, 235) observed that "an attempt to correct past mistakes in vesting property rights by simple deprivation or confiscation may have only distributional effects (except insofar as insecurity of rights affects incentives of others) but freezing the right to the original use of water has an adverse efficiency effect from which the community as a whole loses."

Similarly in Montana, the Utah constitution prohibits the transfer and sale of water for use in coal slurry pipelines, suggesting that the state's constitutional reformers somehow knew that coal slurry would never provide a highest and best use for water. Indeed, the economics of water use may have proved that use for coal slurry is inefficient, but this is something that market actors could determine. To the extent that restricted uses could compete with other uses, such provisions keep the price of water artificially low, providing a gain to some users at the expense of others.

Legal restrictions have essentially broken apart the foundation for an effective system of water rights that was built in the "lawless West," as noted by Cuzan (1983, 20–21):

> It is evident that the long-term trend of federal policy has been to mobilize financial, administrative, political, constitutional and judicial resources . . . to gain . . . control of western waters . . . The appropriation doctrine has been undermined, water rights have been virtually expropriated and converted into licenses or permits, and control over western waters has been centralized in state and federal governments.

Instead of relying on markets, many water allocation decisions have been turned over to a rent-seeking process that uses valuable resources without guaranteeing efficiency or equity. If we are to avoid a water crisis through a market solution, we must return to the original principles of the appropriation doctrine.[5] Unfortunately, this has been made all the more difficult by the dilution of appropriative rights with principles from the riparian doctrine.

The Ubiquitous Riparian Doctrine

In addition to regulatory restrictions, judicial interpretations injecting common law precedents of riparian rights into western water law also limited the efficacy of the prior appropriation doctrine. Several elements of the riparian doctrine led directly to more public control of water allocation. First, with riparian ownership, the resource is held in common, thus requiring regulations to restrict access to the commons. Second, because diversions necessarily were prejudicial to other riparian owners, farmers and miners sought and obtained "license, grant or prescription" from legislatures and courts.

Though the informal rules from mining camps and irrigation communities had evolved a fairly well-settled doctrine for defining, enforcing, and transferring rights, disputes arose when riparian landowners believed their rights were violated by diversion. For example, in 1853, the California Supreme Court held in *Eddy v. Simpson* that "the owner of land through which a stream flows, merely transmits the water over its surface, having the right to its reasonable use during its passage. The right is not in the corpus of the water, and only continues with its possession."[6] Though the ruling was subsequently overturned, a riparian precedent was established in California at that early date. Courts continued to hold that rights were only usufructuary and that they were lost once the water left the possession of its appropriator. John Clayberg (1902, 97–98) saw how riparian principles were contributing to the evolution of the prior appropriation doctrine:

> There never seemed any doubt in the mind of the court about the true position to be taken, but it is almost amusing to read their statements as to whether the principle announced was in consonance with the common law, or in departure from it, because of the conditions and necessities of the case. In one case the court would say that they did not depart from the common law but found principles there insufficient to sustain their holdings. In another, the doctrine would be announced that the common law was inapplicable, and that the reasons of that law did not exist in California.

The resulting mixture of riparian and prior appropriation doctrine led to a confusion that stifled more complete definition and enforcement of private water rights. As further explained in Chapter 5, this dilution of appropriative rights shifted many water allocation decisions out of the marketplace and into the political process. Another legal doctrine that diluted appropriative water rights and, consequently, the viability of market-based water allocation, is the public trust doctrine.

The Public Trust Doctrine

The public trust doctrine is a common law principle inherited from Great Britain, originally formulated to protect the public's interest in commerce, fishing, and navigation on navigable waterways. During the twentieth century, the doctrine was expanded beyond its historical application to influence the management of numerous environmental resources including water. The doctrine has proven particularly effective at undermining private water transactions by challenging the validity of the underlying private water rights.

In the 1983 Mono Lake case in California, for example, the California Supreme Court found that the state had a trust responsibility to protect the environment for the people.[7] In that case, water diverted by the city of Los Angeles was dewatering Mono Lake enough to change the aquatic and bird life of the lake. Although Los Angeles was entitled to the water under appropriative rights granted by the state in 1940, the court ruled that the state had to reconsider its decision granting the rights in light of its responsibility to protect public trust values, expanded to include environmental quality.[8]

In 1989, the court enjoined the city from diverting water until lake levels rose 2 feet, and then limited diversions to between 4,500 and 16,000 acre-feet per year until lake levels rose 16 feet (Marston 1994, 15).[9] Because of the cut in diversions, Los Angeles had to spend an extra $38 million annually to acquire water from other sources. The bottom line is that well-settled property rights have been abrogated in the name of an expanded public trust doctrine. If the public trust doctrine is permitted to extend beyond commerce, navigation, and fishing to the environment and trump prior appropriation water rights, there are almost no limits on how states can regulate water use and water marketing. In the words of David Slade (2008, 219), "the clash, and confluence between [the two doctrines] over vested water rights . . . are a harbinger of what is yet to come."

The problem with this litigious approach to allocating water is not only that it is hugely expensive, but that it weakens property rights and the ability of markets to promote investment, trade, and the efficient use of water. The public trust doctrine, which can be applied retrospectively to roll back preexisting water rights whenever they appear inconsistent with the public trust, is so vague and elastic that it can lead to extensive government regulation of water rights.

Moving Water Policy to Washington

While the public trust doctrine casts a cloud of state control over every prior appropriation right, the commerce clause of the U.S. Constitution raises the specter of federal regulation over all navigable waters. Despite the apparent limitation, the federal government's commerce clause power extends beyond the regulation of water for navigation.[10] Any action affecting interstate commerce on the nation's navigable waterways is subject to federal regulation.[11] A power known as the navigation servitude,[12] for example, empowered the federal government to ensure unimpeded navigation on navigable waterways and to remove obstructions or improve navigable waterways below the high waterline without compensating the adjacent landowner.

Whether waterways were subject to federal regulation under the commerce clause depends on whether the waterway is navigable, and navigable waterways are defined as those susceptible of being used in their ordinary and natural condition as highways for commerce over which trade and travel could be conducted.[13] Because the navigability test is so broad, a large number of streams and rivers are subject to federal regulation and the navigation servitude (Goplerud 1995, 8).[14] These expansive federal powers over navigable waters effectively prevented private parties from developing reservoirs or rivers. In 1957, Charles Corker (616–17) noted the implications of the federal government's power over navigation:

> The Congress and the courts have been content to treat the word "navigation" as an open sesame to constitutionality. So long as Congress uses the word in statute and the case relates to something moist, the Court takes at face value the declaration that the legislation is in furtherance of navigation. Moreover, the tests of what constitutes a navigable stream has been stretched to embrace most of the waters of the United States.

The early applications of navigation servitude did not foretell the extent to which the doctrine would increase the federal government's potential to regulate water use. In April 2009, the Clean Water Restoration Act was introduced in the United States Senate as a way of amending the Federal Water Pollution Control Act (commonly known as the Clean Water Act), replacing the term "navigable waters" with "waters of the United States."[15] In effect, the measure would give the federal government jurisdiction over all waters, whether navigable or not, including "waters subject to the ebb and flow of the tide, the territorial seas, and all interstate and intrastate waters and their tributaries, including lakes, rivers, streams (including intermittent streams), mudflats, sandflats, wetlands, sloughs, prairie potholes, wet meadows, playa lakes, natural ponds, and all impoundments of the foregoing." If passed, landowners and businesses could face new regulations, permitting requirements, and potential restrictions on land and water use. In other words, the amendment would expand central control over resources at the expense of private property rights.

The federal government's reach into western water rights was also expanded by the federal reserve water doctrine under the property clause of the Constitution. According to this doctrine, retention of public domain by the federal government for Indian reservations, national parks, and national forests implicitly included the reservation of water, even if unappropriated, to fulfill the purposes for which the land was reserved (Sax and Abrams 1986, 494). Unfortunately, most of the reserved rights have not been quantified, making conflict with private appropriation rights inevitable.

To understand how this doctrine can cloud water rights, consider the context in which the reserve water doctrine evolved. The federal government established the Fort Belknap Reservation in Montana in 1888, but did not quantify any water rights and did not divert water for irrigation, the main criterion for claiming an appropriative water right. Subsequently, non-Indian settlers came to the area and developed irrigation systems, appropriating what they considered unappropriated water from the Milk River. When the Indians began diverting water in competition with the settlers' diversions, the settlers built dams and reservoirs upstream of the reservation to ensure irrigation of non-Indian lands. These projects in turn reduced water availability for the reservation, prompting the U.S. government to sue the settlers as trustee for the Indians. Eventually, the court, in a decision that became known as the "Winters Doctrine"[16] (named for the judge), enjoined the settlers from diverting any waters that interfered with Indian water rights, finding that "when the Indians made the treaty granting rights to the United States, they reserved the right to use the water of the Milk River, at least to an extent reasonably necessary to irrigate their lands."[17] Rodney Smith (1992, 169) summarizes the ambiguities that ensued:

> The *Winters* decision set the parameters for Indian water rights litigation in terms of the amount of water "reasonably" necessary for the purposes of the reservation. In assessing what constituted reasonably necessary, the court would not restrict itself to current uses but would also consider "future requirements." Without specific criteria for determining what constitutes "reasonably necessary" and "future requirements," the court raised the specter of complex factual inquiry into the economics of water use on and off the reservation and suggested possible consideration of the equitable interests of the parties.

There has since been a cloud over the quantity and security of appropriative water rights when federal reserved water rights are involved. To make matters worse, the federal government has begun claiming reserved water for wilderness areas and national parks, where it is even more difficult to quantify how much water was implicitly reserved.

Since its inception, the prior appropriation doctrine has undergone a series of dilutions, restrictions, rewrites, and curtailments. All of these have undermined the doctrine's capacity to foster water marketing in the West. But perhaps no other force has crowded out or complicated the private water market more so than federal reclamation.

Water Gushes Uphill to Politics

When the Bureau of Reclamation entered the water storage and delivery business in 1902, the potential for water marketing was further undermined. The bureau began telling water users where, when, and how they could use the water from federal projects. Because government money was being used to provide the water, politicians and bureaucrats claimed that they had a right to dictate water allocation.

The initial Reclamation Act of 1902 required that construction funds for dams and delivery systems be repaid within 10 years, thereby limiting the period of federal control. But as the rent-seeking process encouraged the extension of repayment periods to over 50 years, federal control was expanded and strengthened.

The vagueness of the reclamation statues regarding the transferability of federal project water coupled with the laws' restrictions, such as the appurtenancy requirement and acreage limitation, also fostered bureau control. In addition, the federal government obtained state water rights for project water and then entered into contracts with irrigation districts that in turn delivered the water to its final users, creating uncertainty about whether districts or individual users had the authority to transfer it.

Decades ago, Charles Meyers and Richard Posner (1971, 20) pointed out the extent of the Bureau of Reclamation's control over transfers.

> Even after repayment of its loans, the bureau, it appears, retains title to the dams and reservoirs constructed under the project and with them title to project water. In addition, many pay-out projects have rehabilitation contracts with the bureau, which give it a continuing interest in the financial integrity of the project. The bureau's interest in projects that have not paid out is clear, since it looks to the individual farmers or to the district . . . for recovery of the costs allocated to irrigation. At all events, whether by statute, expressed or implied contract, general understanding based on past and expected future favors, or some combination of these, the bureau's consent must be obtained for the transfer of any significant quantity of water, supplied by either a paid-out or nonpaid-out project, where the transfer involves either use on a different parcel of land or a different use; and this is so whether the transfer is within or outside project boundaries.
>
> (Meyers and Posner 1971, 20)

The bureau's control over water transfers creates tremendous potential for rent seeking. Bureaucrats can retain budget and power through their authority to grant or restrict transfers. Water users have an incentive to expend effort and resources in attempts to influence bureau decisions because they will affect their wealth positions.[18]

In the 1980s, the Bureau of Reclamation began to re-characterize its role in Western water from dam-building to water management. To that end, in 1988 the Department of Interior (of which the Bureau of Reclamation is a part) published principles governing voluntary transfers of reclamation project water. Criteria and guidance for application of the principles followed in March 1989. The purpose of the principles was to clarify the Department of Interior's policy governing water transfers and its intent to facilitate them.[19] The principles:

1. acknowledged the primacy of state water law in allocation and management decisions;

2. outlined when the Department of the Interior would become involved in transactions for reasons such as federal contracts, federal water rights, legal obligations, or requests for involvement by non-federal entities;

3. conditioned participation in or approval of transfers on adequate protection of third-party consequences[20] and on maintenance of the federal government's financial, operational, and contractual position following the transaction;

4. limited the department's participation to transactions that are in accordance with state and federal law and are proposed by others;

5. considered irrelevant the fact that water involved in a transaction has been federally subsidized;

6. expressed an intent to allow transferors of project water to realize a profit without the bureau claiming some of that profit;

7. stated its commitment to comply with the National Environmental Policy Act and to mitigate any adverse environmental effects arising from proposed transactions.

(U.S. Dept. of the Interior, 1988)

Although the Department of the Interior's principles were the first steps in the development of a federal policy favoring voluntary transfers of project water (see Smith and Vaughan 1991, 13), the reclamation statutes themselves remain the greatest restriction on federal water because they limit the uses to which project water may be applied and fail to establish adequate procedures for water transfers.

In 1992, Congress took a step toward encouraging transfers when it passed the Central Valley Project Improvement Act (CVPIA).[21] That legislation authorized all individuals or water districts receiving Central Valley Project water under service, repayment, or exchange contracts to transfer all or a portion of that water to other California water users.[22] While the act apparently encouraged water markets, it effectively nixed trades by imposing 13 conditions on transfers, by favoring fish and wildlife interests at the expense of all other interests receiving CVP water under water contracts,[23] and by giving state and federal regulatory agencies power to achieve a "reasonable balance" among competing demands for CVP water. Such restrictions impede, rather than promote, efficient market allocation (Gardner and Warner 1994). The onus, therefore, still remains on Congress to revise reclamation legislation to facilitate markets in order to give teeth to any federal policy favoring transfers.[24]

Conclusion

Between regulation and reclamation, the market has been crowded out of many water allocation decisions. In most western states, water is the declared property of the state, the people, or the public. Only in Colorado and New Mexico is this declaration limited to unappropriated water. Writing more than 50 years ago, Hirshleifer et al. (1960, 249) concluded that

> the current trend . . . runs strongly against the development of a system of water law based on individual choice and the market mechanism. . . . the evidence is fairly clear that the tenor of the legislative and judicial edicts . . . is the product of the ignorance of even importantly placed and generally well-informed individuals today about the functioning of economic systems—and, in particular, it is the product of the common though incorrect opinion that the public interest can be served only by political as opposed to market allocation processes. . . . That there are defects in the present systems of private water rights is very clear; but to abolish property rights rather than cure the defects is a drastic and, we believe, unwise remedy.

With few exceptions, legislative and judicial actions have continued to erode the basis of private property rights in water. Lasky's concern in 1929 over the shift from prior appropriation to economic distribution of water by the state was certainly prophetic (Lasky 1929, 171). In 1991, law professor Charles Wilkinson (1991, v) eulogized the death of the prior appropriation doctrine stating "Prior Appropriation passed away last month at the age of 152. Prior was a grand man and led a grand life—by any standard he was one of the most influential people in the history of the American West."

Federal reclamation has also constrained the ability of markets to allocate water efficiently.[25] The federal government is currently the largest wholesaler of water in the country, providing roughly half of the surface water withdrawn for irrigation (U.S. CBO 2006, 5). The growth of federal water projects and the Bureau of Reclamation have displaced water marketing opportunities and promoted inefficient water allocation and use at the expense of hydropower consumers and taxpayers. Fortunately, despite widespread plans for continued expansion of federal water projects, there is growing pressure to change federal policies to address waste and improve allocation. These objectives can be met through market mechanisms. In a 2006 report, the CBO stated that the federal government could "encourage efficient water use" by "reconsider[ing] existing subsidies for water delivery and agricultural products" that currently distort price signals and "impede the transfer of water resources to higher value uses." Moreover, the expansion of water marketing opportunities and the use of water banks would "facilitate the legal transfer and market exchange of various types of entitlements for surface water, groundwater, and storage" (U.S. CBO 2006, 13).

Realigning state and federal policies to facilitate improved water allocation and incentives for use efficiencies will not be easy. However, maintaining the status quo in an environment of increasingly scarce water resources and growing demands will at best perpetuate massive subsidies, rent-seeking behavior, and the threat of water crises. Reclamation may have made the desert bloom, but there is little economic justification for the blossoms.

5

BACK TO THE FUTURE

From the western frontier, especially the mining camps, arose the doctrine of prior appropriation and a foundation for water marketing. Also known as the Arid Region Doctrine, this institution governed the ownership and use of water based on "the recognition of discovery, followed by prior appropriation, as the inception of a possessor's title, and the development by working the claim as the condition of its retention" (Kinney 1912, 598). This system provided the essential ingredients for a market in water because it defined and enforced water rights and allowed them to be traded. With the rights to water clearly assigned, water owners had to consider the cost of using the water vis-à-vis what it was worth in alternative uses. If the latter value was higher, there was an incentive to reallocate water to the higher valued uses.

As water was reallocated from one use to another, there was always the possibility that transfers might violate the rights of other users. For example, if upstream diversions were switched to a use that consumed more water and thus left less water flowing downstream or if a change of use dumped effluent into the stream, downstream users would be affected. This potential harm led to the "no injury" rule that all appropriators—senior and junior—were entitled to the continuation of stream conditions as they were when appropriation began.[1]

To protect the rights of downstream users, legislatures, courts, and administrative agencies intervened in the transfer process by requiring water transferors to prove that they were not impairing other users. Since the frontier days when the number of users and trades was small, the potential for conflicting claims on water has increased and with it, government oversight has played an increasing role in market transfers. In some cases, restrictions have protected other water rights owners. For example, state agencies routinely require that proposed transfers go through a hearing process in which potentially harmed water owners can protest changes. If other water owners can prove harm, agencies disallow transfers.

In other instances, government intervention in water transfers has allowed special interest groups to redistribute water's bounty. Perhaps the clearest example of this is the application of the public trust doctrine which has given non-rights holders the ability to block water reallocation[2] and recreationists the ability to access streams flowing through private property.[3] The public trust doctrine is by no means the only restriction on the prior appropriation doctrine's ability to foster water marketing. Other restrictions imposed on the prior appropriation doctrine include the beneficial-use standard, diversion requirements, inability of appropriators to keep or transfer conserved water, the "use it or lose it" rule, and the no-injury rule. This chapter outlines the motives and consequences of each restriction, then explores whether the prior appropriation doctrine can be salvaged.

Constraints on Prior Appropriation

Allegations of market failure have supported numerous restrictions on water marketing. Concerns about monopoly distortion, imperfect capital markets, and third-party effects were used to justify the divestiture of rights and dilution of private markets, as discussed in Chapter 4. Because the prior appropriation doctrine was fundamental to market-based water allocation, it elicited a particularly negative reaction.

According to Trelease (1976, 284), the three "recurrent reactions" to the prior appropriation system were:

1. A dislike of the "property system": appropriators seize valuable interests in the public domain and enrich themselves at the expense of the public.
2. A mistrust of the "market system": a fear that under prior appropriation, water rights will become "frozen in the pioneer patterns," unsuitable for modern times and problems, and not just to reallocation to new uses and needs.
3. A dislike of the "priority system": in a shortage, an "all-or-nothing" rule gives one of two essentially similarly situated water users all of his water while his neighbor gets none.

All of these objections to water rights and water markets are based on a failure to appreciate the flexibility and variety that markets can provide to water allocation. Fundamental to this flexibility, however, is the security of right that originates from a property system, the transferability of right possible under a market system, and the clarity of right created by the priority system. This section details the specific restrictions that undermined the security, transferability, and clarity of appropriative rights.

Beneficial Use

Beneficial-use restrictions are perhaps the oldest in the system of water rights, dating back to English common law. These restrictions found their way into the early mining camps in the West where the appropriators themselves required that water be put to a beneficial use or lost to other users. Beneficial use can be justified on the grounds that it was a low cost way of proving an ownership claim. Today, however, limiting the definition of beneficial use and requiring a diversion can thwart efficient reallocation, especially in the case of instream flows.

The ruling in an 1897 Nevada court case, *Union Mill and Mining Co. v. Dangberg*,[4] was typical of early decisions holding that "under the principles of prior appropriation, the law is well settled that the right to water flowing in the public streams may be acquired by an actual appropriation of water for a beneficial use" (quoted in Tregarthen 1977, 144). The list of uses generally considered beneficial included mining, domestic use, stock watering and later, irrigation.

In the early mining camps, the beneficial-use standard made sense because it provided a way of defining water rights and eliminated the need for a complicated recordation system. By defining the right according to use, it was clear who had water rights. The beneficial-use criterion also encouraged application of water and discouraged speculation (acquiring a water right and leaving the water in the stream for future sale or use). If it were enough to claim water without putting it to use, individuals could have claimed entire watersheds by staking their claim at the mouth of a stream or river and thereby extracted monopoly profits from anyone hoping to use water upstream.[5]

Today, however, the beneficial-use requirement, coupled with the "use it or lose it" principle, can promote the wasteful application of water. It locks water into historical uses that were considered beneficial when they were initiated but that now have relatively low economic value. For example, irrigating hay meadows may have been the highest valued use of water in the late nineteenth century but, today, using the same water to provide instream flows for fish may be more valuable. If acquiring diversion rights and leaving them instream is not a beneficial use, then individuals or groups attempting to move water from the lower to the higher valued use may lose the right due to abandonment. In this way, requiring users to continue diverting their water for strictly defined beneficial uses can prevent gains from trade and stimulate conflict.

Beneficial-use restrictions also encourage individuals to use the law to preclude competing uses. If a state is to enforce use restrictions, it must specify what constitutes beneficial uses. Most western states do so in their constitutions or laws, defining beneficial use as "a use for the benefit of the appropriator, other persons, or the public," and listing certain uses as examples.[6] Such lists now typically include municipal, industrial, and hydropower uses along with domestic use, irrigation, mining, and, more recently, environmental amenities such as instream flows. In some cases, states have specifically excluded certain uses from being considered beneficial. For example, between 1979 and 1985, a Montana statute stated that use of water in coal slurry pipelines could not be considered a beneficial use.[7]

Historically, instream uses such as recreation and fish and wildlife were generally not listed as beneficial uses. For example, the Colorado Supreme Court ruled in 1965 that there was "no support in the law of this state for the proposition that a minimum flow of water may be appropriated in a natural stream for piscatorial purposes without diversion of any portion of the water 'appropriated' from the natural course of the stream."[8] Nowadays, all western states recognize certain uses of instream flows as beneficial, but some states define beneficial instream uses quite narrowly. For example, Wyoming limits instream flows rights to the "minimum flow necessary" to "establish or maintain fisheries" (if the water is supplied from stored sources) and to "maintain or improve existing fisheries" if appropriated from unappropriated waters.[9] In contrast, instream flows in Washington are deemed beneficial if used for fish and wildlife maintenance and enhancement, protection of game and birds, recreation, scenic, aesthetic and all other uses compatible with the enjoyment of the public waters of the state.[10]

Beneficial-use requirements hamper water markets by obstructing the movement of water to its most valued uses. Apart from beneficial-use restrictions and the diversion requirement, instream flows would likely have become an accepted use of water long before now. As long as beneficial use is determined in legislative, judicial, and administrative forums, a great deal of time, effort, and money will be devoted to the governmental process. The "doctrine of beneficial use, with its implications of judicial determination of need and non-use in effect increases the uncertainty of title to rights in water, and therefore reduces their marketability" (Tregarthen 1977, 145).

Rather than lifting beneficial-use restrictions, some scholars advocate expanding the concept in the name of the public interest at the expense of private water rights. According to David Getches,

> All private water rights are conditioned on their being dedicated to a beneficial use. This principle provides a legal basis for exercising the state's police power and enacting regulations to ensure that existing rights satisfy the condition. Thus, even the oldest water rights can be regulated to protect the paramount interests in water asserted by the public. This regulatory purpose justifies limiting water uses if they interfere with water quality or jeopardize publicly important purposes like fishing and recreation. . . . Applying property law principles developed in land use cases, a court could allow governments to regulate water rights virtually without limit if necessary to ensure that they are used in accordance with contemporary notions of what constitutes a beneficial use.
>
> (Getches 1993, 128–29)

Use It or Lose It

Historically, an especially problematic restriction in most western states is the requirement that water must be diverted from the watercourse otherwise the water right will be lost to its original owner and will become available to other appropriators. States have since relaxed this restriction through provisions for the protection of instream flows; however, the opportunities for individual landowners to leave water instream are often limited.[11] Most states prohibit users from leaving water instream for speculative purposes or when not in use, sometimes referred to as "parking" water.

As discussed above, this requirement provided an efficient means for a diverter to establish his claim to the water and to specify the location of the water right (Clark et al. 1972, §409.2). Now that most streams are fully appropriated, rights are well enough defined that the diversion requirement is no longer necessary. Nonetheless, diversion requirements remain a part of state water rights systems, and where they are strictly enforced, they discourage conservation because water rights holders fear that they must "use it or lose it."

The "No Injury" Rule

When water is diverted by an upstream user, that diversion can affect the rights of downstream users either because there is a return flow that is claimed by others or because there is a change in water quality. If downstream claims are junior to upstream claims, the downstream user must take the quantity and quality of the return flow as given. However, if an upstream user changes the point, the time, or the place of diversion or changes the use of the diversion, the quantity or quality of the return flow may change and other water users on the stream could be harmed. This potential harm is known as third-party impairment. As the prior appropriation system evolved, the law considered these third-party effects but generally restricted claims of harm to water rights owners directly impacted by changes of use.

The possibility of third-party impairment prompted all western states to implement judicial or administrative procedures that must be followed before water use can be altered or water rights transferred. Although these procedures vary from state to state, they typically allow water-use changes or water rights transfers only if there is no injury to other water rights holders. This standard is known as the "no injury rule" (Thompson 1993, 701). In Alaska, Arizona, California, Idaho, Kansas, Montana, Nebraska, New Mexico, North Dakota, Oklahoma, Oregon, South Dakota, Texas, Utah, Washington, and Wyoming, administrative agencies are empowered by law to approve or disapprove changes. In Colorado, water courts determine whether changes in use or ownership should be allowed, but the same basic procedure is followed. If the owner of a water right wants to change the point of diversion or the method or place of use, he or she must petition the appropriate state agency or the water court for approval. The agency then studies the proposal to determine whether third-party effects are involved. Notice of the proposed change is published in a regional newspaper to inform other users on the stream system. Objections to the change may be filed, but generally objections are limited to only those who hold water rights.[12]

Most often disputes are resolved privately between the parties involved or by hearing before the agency or water court. Transfer applications may be approved or denied as filed, but are usually approved subject to conditions that protect the interests of the objectors (Gould 1989, 464). Those disgruntled with the transfer decision may appeal either within the agency or to the state's courts.

If third-party impairment is a violation of the property (water) rights held by others, it is a problem of definition and enforcement that must be addressed when dealing with water transfers. The question is which process best solves the problem. Historically, states have dealt with third-party impairment through administrative hearings wherein potentially harmed parties can contest transfers. Montana state law is representative. It allows changes in appropriative water rights when the appropriator proves by a preponderance of the evidence that the proposed new use will not adversely affect the water rights of other persons and the proposed use is a beneficial use.[13] This process can increase transaction costs by delaying transfers, but such transaction costs are part and parcel of a market where the property rights of others are protected. However, if the transaction costs are driven up by claims of impairment by people who actually have no legitimate impairment claim, then the process may artificially raise transaction costs and thwart water markets, as we shall see below.

Area of Origin Restrictions

Restrictions on transfers tend to go beyond the traditional definition of third-party impairment to include arm's length impacts of water transfers. Accordingly, third-party impairment has been applied to environmental impacts, local economies and cultures, and the "public interest" in general (Thompson 1993, 704, 708), where third parties claim impairment without owning water rights.[14] When regulations are expanded beyond traditional water rights to include more distant third parties, they enter the realm of rent seeking to protect status quo wealth from water reallocation that can improve efficiency.

Area of origin protection laws that limit transfers to other basins offer an excellent example of rent-seeking regulation in the name of third-party impairment. These laws may totally prohibit inter-basin transfers or limit them by imposing a tax on transfers as a type of impact fee or requiring payment of compensation to third parties in the area of origin. California allows an originating basin to recall the transferred water in the future if needed (Colby 1988, 738–41; Reisner and Bates 1993, 70–73). New Mexico transfer statutes require a reservation of a share of the originating basin's water supply, but do not clearly define how large the share should be or under what circumstances it can be recalled (Colby et al. 1989, 713). Oregon requires a reservation of 25 percent to the state of any conserved water that is transferred out of a basin (Willey 1992, 403).

Area of origin laws are directed at people who interact with water owners rather than at water owners themselves. In light of the fact that approximately 80 percent of western water is used in irrigation, transfers of water from irrigation to other uses could erode rural county tax bases, shut down businesses dependent upon agriculture, impact the local labor market, and threaten the social values and culture that have developed around the agricultural community.

Not surprisingly, people with these interests support legislation that prevents changes in water use, claiming third-party impairment. California's Water

Resources Control Board can deny a proposed water transfer if it would "unreasonably affect the overall economy of the area from which the water is being transferred."[15] Similarly, in Idaho, applications to transfer water out of agricultural uses may be denied if the change "would significantly affect the agricultural base of the local area."[16] In Colorado, water transfers may be subject to annual payments to the state and local communities to mitigate lost tax revenues from taking lands out of agricultural production (fallowed lands are taxed at a lower rate).[17]

To be sure, third parties feel the impact of efficiency-enhancing water transfers, but this is always the case with market transactions. If a new restaurant opens that provides better service and food and attracts customers from existing restaurants, an argument can be made that existing restaurants are third parties who have been harmed by the competition. This impact is part of the market process that provides consumers with better products and improves resource allocation. Similarly, if a water owner sells his or her water to another user who values it more highly, efficiency is improved though others in the marketplace may be harmed. Naturally these third parties would like to avoid the harm, and therefore they seek legal restrictions such as area of origin legislation. The protection in this case, however, is very different from the protection of water rights owners with prior appropriation claims to not have their rightful claims disrupted. Legislature and courts can create new rights for special interest groups such as those mentioned above and, in so doing, open the door for rent seeking and increased transaction costs that limit efficiency gains.

To legitimize claims in the rent-seeking process, claimants tend to exaggerate economic, cultural, and environmental impacts. However, transferring water from agricultural to urban uses need not dry up and shut down rural areas. In fact, such transfers can serve to strengthen the economies of rural areas. For example, the diversion requirement, use-it-or-lose-it rule, and government subsidies give irrigators the incentive to apply their water to marginally productive lands and crops. According to Wahl (1989, 188–90), reduced production as a result of water markets is likely to be relatively small and limited to these least productive lands.[18] Agricultural water use could be reduced between 15 and 20 percent through conservation measures without significant decreases in production.[19] Similarly, research in California revealed the effects on local economies were small when as much as 29 percent of the land was fallowed (Public Policy Institute of California 2003). Thus, water markets could actually increase the economic stability of rural areas by encouraging improved agricultural efficiency and increasing incomes.

Wahl also makes a crucial point about prices and demand. Agricultural sector representatives often raise the argument that because cities can pay so much for water, municipal demands will put agriculture out of business if markets are allowed to operate. In response, Wahl states:

> As with most any commodity, the quantity consumers are willing to purchase is a decreasing function of price. Although growing municipalities are willing to pay high prices for some additional water, often hundreds of dollars per acre-foot, they would not be willing to pay similar prices for all of the great

quantities of water currently used in agricultural production. Also, as cities buy successive quantities of water from farms, the price for obtaining additional amounts will rise. Eventually, a point of equilibrium will be reached at which additional purchases become unattractive. In other words, water has its value in food production also, and, even though some water will be bid away, not all will be.

(Wahl 1989, 190)

There is little support for the claim that out-of-basin water transfers will deprive rural areas of cultural and environmental values (see Libecap 2005). According to the 1987 report by the Western Governors' Association Water Efficiency Working Group, "It does not appear, westwide, that water markets present the threat to traditional lifestyles or natural areas that is feared. There will just not be that much demand for water to result in wholesale abandonment of rural areas or wholesale destruction of natural areas" (Western Governors 1987, 110). Hanak's (2003, 81) survey of the literature points out that effects of fallowing irrigated farmland are likely to have no more than a 1 percent effect on overall county economic activity, even when payments for economic adjustments are not included.

The Public Interest

Requiring consideration of the "public interest" in state water permitting and transfer procedures is another method of rent-seeking that goes beyond protecting third-party water rights holders.[20] All western states except Colorado and Montana require that water officials consider the public interest in issuing new permits to appropriate water (Gould 1989, 473), but many do not require such consideration with water transfers (Ingram and Oggins 1992, 516). Nonetheless, the push toward regulatory consideration of the public interest is increasing. Because the public interest is never specifically defined, its use as a rent-seeking tool can be expanded to include almost any special interest under the guise of community values (economic, cultural, and social), environment impacts (wetlands and other ecosystems, fish and wildlife, and water quality), and recreational amenities, to mention a few (Gould 1989, 446–48).

Bates et al. (1993) advocate water markets, but in a role secondary to advancing the public interest.

A hard look at water policy should seek distributional fairness. . . . The public, through some acceptable process, must first decide which waters are for public use and which are available for private use within a market system. . . . [Private] appropriation ought to be limited to the amount that is not needed by the whole community for the satisfaction of public values.

(Bates et al. 1993, 185).

Bates et al. (1993, 186) fall short of calling for water to be public property and instead suggest that "the use of waters appropriated for private use is subject to regulation like any other activity. For instance, regulation can limit pollution and define how markets operate . . . State and federal agencies can further the principle of fairness in water decision making by creating new processes for comprehensive, integrated decisions . . ."

Perhaps the most pernicious tool for limiting water rights in the name of public interest is the public trust doctrine. This legal doctrine came to the United States from English common law, where it served to protect the right of the public to use navigable waterways for transit, commerce, and fishing (Jaunich 1994). In recent years, however, the doctrine has been broadened by federal and state courts to protect a range of public interests far beyond navigable waterways, including instream flows for fish and wildlife, recreation and scenic values (Huffman 2008). Application of the public trust doctrine can trump private property rights by reallocating water to public uses without compensating the rights holder as in the Mono Lake case.[21] In this way, the doctrine has become a powerful tool for environmental organizations to influence water allocation decisions through the court system.

A recent example of public trust litigation involves the pumping of water from the Sacramento–San Joaquin River Delta to Kern County, California. In September 2010, the California Water Impact Network, the California Sportfishing Protection Alliance and AquAlliance filed suit against two state agencies alleging their violation of the public trust doctrine. The lawsuit claims the State Water Resources Control Board failed to enforce permit and licensing conditions on the California Department of Water Resources' operation of state pumping stations. In particular, the complaint alleges both agencies violated the state's public trust law by failing to consider the environmental impact of increasing annual pumping out of the delta. As a consequence, according to the complaint, the pumping operations have caused extensive damage to the Bay-Delta estuary and its fish and wildlife populations.

If the plaintiffs are successful, historic water users in the delta could lose all or a portion of their water rights and may be forced to forgo diversions during critical growing periods. This possibility undermines the security of these users' water rights and casts significant doubt over the financial viability of the region's agricultural operations. Moreover, regardless of the claim's validity, negotiation could provide the parties with a cheaper and more immediate resolution.

Limitations on Conserved Water

Rules prohibiting appropriators from keeping or transferring water conserved or salvaged by lining ditches, installing drip irrigation systems, or changing crops also discourage conservation. The restrictions are based on the belief that any conserved water must have been previously wasted and not put to a beneficial use. Hence, the wasted water belongs to the watercourse from which it was originally diverted.[22] By not allowing appropriators to enjoy the fruit of their efforts, any incentive to conserve water through curtailing wasteful practices is eliminated (Clyde 1989, 442). Unfortunately these restrictions remain in effect in many states and reduce the incentive to employ water conserving measures.

Preferential Uses

Legislative prioritization of uses is another restriction on appropriative water rights that limits water marketing. Early rulings by the California Supreme Court "asserted that 'a comparison of the value of conflicting [water] rights would be a novel mode of determining their legal superiority.' Thus, it became a fundamental axiom that each of the purposes 'to which water is applied . . . stands on the same footing'" (quoted in McCurdy 1976, 257). In 1876, the Colorado state constitution declared that when water is used for the same purpose, priority and time shall determine the superior right, but it also established a hierarchy among different uses "when the waters of a stream are not sufficient for all desiring its use." In this case, preferred status is granted to domestic use over agriculture and agriculture over manufacturing (Trelease 1955, 134).

Most western states have followed Colorado's precedent by including preferred uses in their constitutions or legal codes. Although there is a wide variation in preferences, domestic and municipal uses are generally preferred over agricultural, industrial, and instream uses. Despite the appearance that preferences would require reallocation to preferred uses in times of shortage, courts have rarely interpreted them in that way because it would upset the prior appropriation system that is based on first-in-time, first-in-right. Some courts have even held that application of preferences would be a taking of property requiring compensation. All western states allow condemnation of water rights for municipal use and many require eminent domain condemnation and compensation to effectuate preferential uses (Clark et al. 1972, §408.4; Getches 1984, 108–10).[23]

In a market setting, if a water owner values his water more highly than does a competitor, market forces would ensure that he would not sell and that the water would remain in its highest valued use. With preferred uses and the power to condemn, however, competitors can obtain legally defined "inferior" water rights provided they pay "just compensation." In this setting, the value of the condemned water right and hence the compensation is determined by the judicial process where opportunity costs may be ignored. In the absence of mutual consent, there might not be gains from trade or efficiency.

Fortunately, municipalities seldom exercise their condemnation powers in this manner. Nevertheless, preferential use statutes have worked hand-in-hand with beneficial use and the other restrictions discussed above to reduce the security of water rights, thus reducing the potential for water marketing and encouraging rent seeking. Though these restrictions on water rights have been justified in the name of market failure, they have actually served to encourage market failure by weakening water rights established under the doctrine of prior appropriation and preventing markets from improving water-use efficiency.

For the market to allocate water resources, rights to water must be well established and transferable. Modifications to the prior appropriation doctrine have interfered with both. For example, in order to obtain a permit to use water, an individual must demonstrate that the water will be put to a beneficial use, the determination of which is left to the political and judicial processes. Market criteria are not the determining factors. Restrictions are placed on intra- and inter-basin transfers of rights as well as on changes of use, encouraging rent seeking and distorting the true opportunity costs of water. Moreover, it is not always clear whether water from federal reclamation projects may be legally transferred in time to meet demands. As a result of all of these restrictions, markets may fail to allocate water efficiently but the fault lies with politics, not markets.

Simple Changes, Big Results

To understand how some simple changes could improve the efficiency of water markets, consider the following hypothetical example (see Figure 5.1). The Gallison River is being used to supply water for municipal, agricultural, industrial, and recreational uses. The river is located in a western state where the doctrine of prior appropriation evolved from early mining and farming interests. Average annual flow in the Gallison is 2,000 cfs.[24]

First settlement along the river was in 1862 when Farmer Leal began supplying agricultural products to nearby mining camps. At that time, he constructed an irrigation canal that diverted 500 cfs of water from the river. The crops he planted and the irrigation techniques he used resulted in a return flow of 250 cfs to the Gallison. In 1872, Farmer Hill settled in the Gallison Valley and constructed his own canal, which withdrew 1,000 cfs and returned 450 cfs. The combination of withdrawals and return flows left 1,200 cfs of water unclaimed. Farmer Meiner arrived in 1877 and claimed 500 cfs, which he diverted through an ingenious canal system into a neighboring basin. Hence, the return flow from his withdrawal was zero.

By the time the town of Waterville was established in 1901, the procedure for establishing water rights had been turned over to the state water resources agency. The community's application for a permit to use 500 cfs for domestic consumption was not challenged. But when a cement plant located in the next basin sought a permit to divert the 200 cfs remaining in the river downstream from Waterville and transfer it out of the Gallison Basin, local residents protested, and the agency refused to grant the permit on the grounds that the transfer could jeopardize future development in Gallison Valley and therefore would violate the state's recently enacted area of origin protection law. Eventually, the 200 cfs were claimed by a downstream user in the next state.

As the growth of Waterville led to increased municipal water demand, Waterville was forced to purchase rights from other users. When Waterville attempted to purchase the 200 cfs from the user in the downstream state, that state's water engineer disapproved the transfer. Waterville then attempted to secure the water from upstream farmers but found them unwilling to sell at the price Waterville offered. Thus, the community exercised its eminent domain power and condemned 200 cfs of Farmer Meiner's water right, obtaining it at the price originally offered.

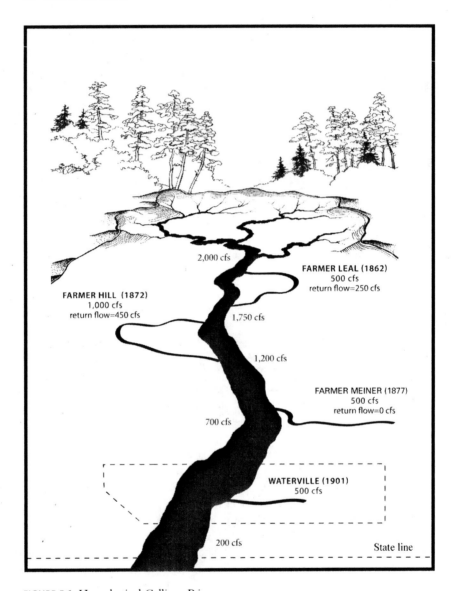

FIGURE 5.1 Hypothetical Gallison River

Farmer Meiners then attempted to purchase 1,000 cfs of water from Farmer Hill to replace the 200 cfs condemned by Waterville to drown gophers in fields he owned in a neighboring basin. The agency denied his transfer application stating that if Farmer Meiners diverted and removed 1,000 cfs out of the Gallison Basin without returning 450 cfs as had Farmer Hill, third parties downstream would be impaired. Moreover, using appropriated water to flood gopher holes was not considered a beneficial use under state law.[25]

To further complicate water rights in the Gallison Valley, Leal, Hill, and Meiners formed the Gallison Conservancy District and successfully transferred their water rights to the district in exchange for transferable shares. Then they lobbied for congressional authorization to construct a Bureau of Reclamation dam on the Gallison River just above Waterville. Since the dam would leave 700 cfs in the river, its construction would cause no third-party impairment. In 1935, Congress approved funding of the dam with the standard provision that the funds be repaid by the conservancy district. A repayment period of 40 years was established in the water service and repayment contract between the bureau and the district, but bad crops and low prices led the bureau to grant several grace periods during which payments were not required. Until repayment was completed (and very likely beyond), the bureau would retain ownership of the water rights and dam works, and would levy annual operation and maintenance charges on the district.

When a sprinkler irrigation system salesman came through the Gallison Valley and demonstrated that the farmers could cut their water consumption in half by installing sprinkler systems, Farmer Hill almost signed on the bottom line. However, after learning that neither he nor the conservancy district would have any right to the water conserved as a result of his increased efficiency, he abandoned the idea as a waste of money.

After several bad years, Farmer Hill decided to sell his shares in the district to Waterville in order to avoid bankruptcy. Although the bureau was willing to approve the transfer, Waterville backed out of the deal because the dam's authorization act was vague concerning transfer of water outside the project area and use of project water for purposes other than irrigation. In addition, the Gallison Valley Conservancy District balked at allowing transfer of the water outside its boundaries because Farmer Hill refused to give the district any share of the profit he would realize from the sale and because it was unclear whether the town would continue paying Hill's share of operation and maintenance costs.[26] The bad economic conditions facing the farmers in the district, however, did lead to another grace period during which repayment was suspended.

The hypothetical Gallison River example illustrates how water transfers are restricted by modifications to the original prior appropriation doctrine. Questions of third-party impacts, environmental harms, and Bureau of Reclamation approval create a nearly impossible gauntlet of regulatory and judicial review for both intra- and inter-basin water transfers. The question is, could these transfers be efficiently handled through the market process without third-party effects?

The cement plant's petition for unclaimed water for transfer to another basin was a signal that it had a use of higher value than any in the Gallison Valley. It is easy to understand, however, why Gallison Valley residents might protest the transfer to keep the water in their basin in case they might want it in the future. By limiting competition from outside sources, water prices will be kept lower for users in the basin. Unfortunately, it was state law, rather than market forces that prevailed, resulting in an inefficient allocation.

Waterville's attempt to obtain additional water through market exchanges implied the town was willing to pay for more water. If a downstream user in another state was willing to sell to Waterville, the implication is that municipal uses had a value in excess of the water's opportunity costs (assuming the political process reflects municipal value). Upstream users were unwilling to sell their water at the price being offered by Waterville, suggesting that the agricultural value was greater than the municipal value. By allowing Waterville to condemn rights, either water was being transferred to a lower valued use or Waterville residents were obtaining their water for a price that was less than they were willing to pay. In the first case, inefficiency would result; in the second, value (rents) would be transferred from farmers to municipal residents.

The state's policy of not allowing water users to keep or sell water conserved or salvaged by increased efficiency removed all incentive for Farmer Hill to install a sprinkler irrigation system. As a result, water that could be applied to other agricultural, municipal, or environmental use and that could be sold to increase Hill's revenue was locked into its current, less efficient use.

Farmer Hill's attempt to sell his shares in the conservancy district also reflected the potential for an efficient water transfer. It is reasonable that the Bureau of Reclamation would want to protect its financial interest in having the construction costs repaid, and approval of the transfer would increase the likelihood of repayment, especially given Farmer Hill's precarious financial position. Moreover, repayment from Waterville would likely be larger than it would be from Farmer Hill due to the fact that under reclamation law, payments for municipal and industrial use of project water bear interest, whereas those from irrigation generally do not (Wahl 1989, 181).[27] Unfortunately, the transfer did not happen in spite of the bureau's support because the vagueness of the reclamation statutes created legal risks that might invalidate the purchase and because of the conservancy district's opposition to the transfer.

What about transfers that alter the return flow and impact downstream users such as Farmer Meiners' endeavor to buy Farmer Hill's 1,000 cfs? Can a case be made that state intervention is necessary when return flows are altered by changes in use or diversion? As discussed earlier in the chapter, harms can result from ill-defined property rights especially when water is used and reused along a stream basin. Any transfer may affect the return flow available to others (Johnson et al. 1981, 273–74). But judicial and administrative frameworks are certainly capable of addressing and even eliminating the impacts of water transfers on third-party users. However, they do so by imposing restrictions and controls. The greater the restrictions, the less likely users will be able to increase water-use efficiency through markets.

While many scholars recognize the importance of allowing water transfers and charging higher prices for water, they simultaneously advocate increased governmental control of transfers as the answer to avoiding third-party impairment (see e.g. Glennon 2009, 309–10). Those calls for protecting third parties by reviewing transfers also expand the list of interested parties beyond immediate water rights holders to include the broader local community, environmentalists, recreationalists, and anyone else even remotely touched by water allocation decisions (see Bates et al. 1993, 182–92). Advocates of area of origin protection, public interest criteria, and the public trust doctrine are motivated by a distrust of markets and the belief that more regulation is necessary to solve problems not addressed by markets. They believe management by disinterested bureaucrats will handle allocation of resources and care for the environment better than markets can (Huffman 1994, 427–28).

To be sure, there are real third-party effects that must be taken into account before transfers are allowed, and administrative reviews offer a mechanism to do this. For example, suppose that Farmer Hill changes his water use by applying it to a new crop that is much more water-intensive. This will alter the return flow from irrigation and potentially impair downstream users whose rights depend on return flows from Hill. Or suppose that Farmer Meiners moves his point of diversion upstream and makes it more difficult for other water users to divert water. In such cases where stream conveyance is affected, third parties have a right to cry foul. Similarly, cases of pollution are examples of third-party impairment that should be guarded against. In all three of these cases, there is a well-identified third party with a clear legal right that has been impaired. Therefore administrative or judicial procedures that limit transfers to prevent harm to third parties are justified on both equity and efficiency grounds.

Extending the web of rights to the general public, however, is a reach with little basis in the common law of torts, nuisance, and trespass and is likely to impede efficiency-enhancing transfers. By granting the public intervention into water transactions, there is no limit to the number of claims of third-party impairment. Such an extension of rights necessarily takes rights from existing owners and transfers them to new potential rights holders. When the environmental groups successfully expanded the public trust doctrine in the Mono Lake case, senior water rights holders lost their ability to divert water and environmentalists gained protection of the things they valued. In finding that the public had a right to use water for protecting environmental amenities, the court opened a Pandora's box that invites other special interest groups to seek water reallocation on the grounds that they are a harmed third party. Not only is rent seeking encouraged in the legal arena, potentially efficiency-enhancing trades are discouraged because parties to trades are not certain of their water rights.

Rather than increasing restrictions on water transfers in order to protect against third-party effects, state transfer processes can work to facilitate water markets. For example, most water rights are specified in terms of the amount of water diverted. It is possible, though not always easy, to determine what proportion of the water is actually consumed—the portion of diverted water that is incorporated into crops, consumed by humans or livestock, evaporated, or transpired. The determination is necessary because return flows are almost always claimed by others who would potentially be harmed if the entire diverted water right were transferred. Once the amount of consumptive use is determined, that amount can be transferred. The determination of consumptive use clarifies water rights and greatly reduces the potential for objections to transfers. Of course, allowing only transfers of consumptive use rights will not solve third-party effects. Problems related to changes in the timing of return flows, changes in the ability of a stream to convey water, or changes in water quality must also be accounted for in the state transfer process.[28] Definition of consumptive use, however, is one means by which state agencies can aid, rather than impede, water markets.

Proof of Concept

Colorado provides an example of what happens to water transactions when water rights are not clear. In Colorado, the owner is not entitled to transfer the quantity of water appropriated but is allowed to transfer only the historic consumptive use amount. In order to determine the transferrable quantity of water from an irrigation use, the courts have focused on the "duty of water," defined by the Colorado Supreme Court as

> that measure of water, which, by careful management and use, without wastage, is reasonably required to be applied to any given tract of land for such period of time as may be adequate to produce therefrom a maximum amount of such crops as ordinarily are grown thereon. It is not a hard and fast unit of measurement, but is variable according to conditions . . .[29]

The problem is determining the consumptive use portion of a water right. The amount of consumptive use is determined on a case-by-case basis following submission of evidence on the transferable quantity of water by all parties involved in the transfer proceeding (Colby 1990, 1191). The adversarial nature of Colorado's transfer proceedings, coupled with the looseness of the definition of the quantity that may actually be transferred, generate significant transaction costs and delays (Colby 1990, 1189). Moreover, the amount of water that may be transacted is ultimately decided by the courts. For example, when the city of Denver sought to purchase water rights whose appropriated total was nearly 400 cubic feet per second, the Colorado court allowed only the transfer of 80 cubic feet per second (Hartman and Seastone 1970, 23–24).

The distinction between diversionary and consumptive use rights in state transfer processes allows for more clearly defined water rights. For example, when the Public Service Company of New Mexico purchased 10.23 acre-feet of water and changed both its location and use, the new water rights were described as follows in the "Dedication of Water Rights" document: "(2) This dedication is for 6.82 acres of irrigated land having a diversion right of 20.46 acre-feet of water per annum and having a consumptive use of 1.5 acre-feet per irrigated acre for a total of 10.23 acre-feet per annum for consumptive use" (quoted in Johnson et al. 1981, 285). Such a specific definition of a water right is common in the New Mexico system.

Other western states have adopted means of streamlining their transfer procedures. Wyoming statutes attempt to reduce the cost of establishing consumptive use rights by establishing a presumption that return flows from irrigation are 50 percent (Gould 1988, 37). Nevada gives objectors to water transfers the option of filing either formal or informal protests. Formal protests require a hearing, while informal protests do not. Nevada also uses a unique method for resolving protests that often results in settlement. The formal field investigation enables parties to meet with the agency official at the site of the proposed change, which lends clarity and understanding of the issues involved in the proposed transfer, as well as less formality to the process as a whole (Colby et al. 1989, 702–3).

Most western states provide for expedited transfer processes in order to meet short-term water needs. In California, transfers of less than a year are generally approved within two months and may be completed in as little as a few hours (State Water Resources Control Board 1999). Short-term transfers are given priority over other transfers and do not generally require compliance with the state's Environmental Quality Act. Similarly, in Oregon, temporary transfers are not subject to the state's formal change of use process that would require a comprehensive evaluation of the water right. Before approving a transfer the state Water Resources Department makes an initial assessment of the potential for third-party injuries, but also reserves the right to revoke the transfer if injury does occur.[30] Generally transfers are in like uses and the risk of injury to other users is minimal. Transfers are approved in an average of 30 days (Donohew 2009).

Water banking, used in Idaho since the 1930s and in California, is another means to streamline water transfers. In Idaho, banking consists of reclamation water users placing unused supplies in the bank on a yearly basis. The bank leases storage space in Bureau of Reclamation facilities, accepts surplus water, and sells it, generally to power companies or other irrigators. The leases are not subject to state transfer procedures because the bureau holds the overall state right to the water within its projects, although state legislation does authorize the water rentals (Wahl 1989, 133–35).

Water banks were used in California in 1977, 1991, 1994, and in 2009 as a response to severe drought conditions (discussed in Chapter 2). The 1977 bank was established by the federal government that spent $2.3 million to purchase 46,438 acre-feet of water from the California State Water Project and from the bureau's Central Valley Project water contractors. This water was then sold to irrigation water buyers experiencing critical shortages. Prices were set so that administrative and transport costs were covered, but no undue benefit or profit was permitted to accrue to the sellers (Wahl 1989, 136–38).

The 1991 water bank was established by the State of California rather than the federal government. It purchased over 800,000 acre-feet of water from users under both the State Water Project and federal reclamation projects at $125 per acre-foot and sold nearly 400,000 acre-feet at $175 per acre-foot.[31] The buyers were mainly municipal and agricultural users.

Although successful in quickly moving water to higher valued uses under drought conditions, these examples of water banking are not perfect examples of water marketing. The Californian water bank failed to facilitate a true market because the state was the only buyer as well as the arbiter of who got the water. The California bank also established the prices at which the water would be bought and sold, rather than allowing the market to determine prices (Hayward 1991, 47). Despite these criticisms, these examples of water banking facilitated short-term transfers during times of drought and encouraged water conservation and efficiency by allowing users with surplus water to lease it to the bank for sale to higher valued users.

Finally, the water transfer processes may be streamlined by emulating special water districts, such as irrigation and conservancy districts and mutual companies, where active markets often exist. As discussed in Chapter 3, water districts are defined geographically, hold water rights under state law, and issue stock to shareholders. Shareholders receive delivery of water from the district on a pro rata basis depending on the number and class of shares held. Delivery of water to shareholders according to the number of stocks held is considered a contractual right, the transfer of which is usually not subject to state laws governing water transfers unless the water rights of the district are affected (Thompson 1993, 712). In addition, because conveyance losses are proportionately shared and return flows are a part of the district's system as a whole, third-party effects are minimal and easily avoided. As a result, transaction costs for transfers are very low, allowing markets to thrive.[32] Moreover, prices are often set by the market to promote efficient allocation and encourage conservation.

The Northern Colorado Water Conservancy District is known for its very active internal water market. About 30 percent of the district's 310,000 shares move through its rental market each year. Prices have varied over the years, reaching close to $3,000 per acre-foot in the late 1970s to more than $12,000 in recent years (Brown 2007). Districts in other states, particularly California, Texas, Arizona, Nevada, and New Mexico also have very active internal markets (Brewer et al. 2006).

Water transfers are not limited to outright sale of water rights or entitlements, but include short- and long-term leases and sale lease-backs—which involve purchase of land and water rights followed by lease of the land and water rights back to the sellers—keeping agricultural lands in production. Cities in New Mexico, Arizona, and Colorado have used sale lease-backs to secure future water supplies.

Subordination agreements are another type of transfer that can be used to benefit environmental and recreational interests. Under subordination agreements, water rights holders with senior priority dates agree to subordinate their senior status to junior rights in dry years, essentially agreeing to share the burden water shortages (Shupe et al. 1989, 420–23). An example of a large-scale subordination agreement is emerging in Colorado where senior rights holders on the state's primarily agricultural West Slope have proposed creating a water bank through which Front Range municipalities with junior rights, such as Denver and Aurora, can acquire water right seniority. The need for such a water bank is driven by the Colorado

River Compact. The Compact is a 1922 multi-state agreement that requires Colorado and the other Upper-Basin states of New Mexico, Utah, and Wyoming to leave 7.5 million acre-feet of water per year, on a 10-year rolling average, in the Colorado River for the Lower-Basin states of California, Nevada, and Arizona. If the 10-year rolling average dips below 7.5 million acre-feet, Lower-Basin states can institute a Compact curtailment, shutting off water rights junior to the 1922 compact.

As the state's population and water demands have grown, meeting the minimum delivery amount has become more difficult and the prospect of a compact curtailment has become a legitimate threat to the water security of Front Range municipalities. By using subordination agreements to transfer seniority to Front Range municipalities and by foregoing their diversions from Colorado River tributaries, West Slope ranchers can reduce the likelihood of a curtailment and the cost to Front Ranger municipalities should a curtailment occur. As an added benefit, the water banking concept also allows environmental organizations interested in increasing stream flows to purchase and retire senior diversion rights. In each case, whether the buyer in the subordination agreement is a thirsty municipality looking to increase water supply security or a conservation organization seeking to improve fish habitat, no physical transfer of water is needed to shift water consumption from lower to higher valued uses.

Conclusion

The incessant conflict that has come to characterize western water management has led some commentators to declare dead the doctrine of prior appropriation (Wilkinson 1991). The cause of the conflict, however, lies not with the prior appropriation doctrine in its original form but in the countless modifications to it which restrict, encumber, or altogether prohibit water users from trading water. While the doctrine of appropriation is not without flaw, the allocation problems in many western states are not so much the fault of a water rights market as much as they are the fault of restrictions placed on appropriative water rights and markets. Administrative agencies and courts continually interfere with the definition and enforcement of water rights and thereby encourage rent seeking.

To salvage the appropriation doctrine and encourage water trading to encourage more efficient water use and discourage conflict, many of the restrictions on water transfers must be removed. When the diversion and use of water cannot be changed, higher valued alternatives are foregone at a cost to both the water right owner and society. If agricultural water users in northern California could transfer their rights or federal contractual entitlements to southern California municipalities, it is likely that water with low marginal value in agriculture would be shifted to the higher marginal valued municipal uses, assuming the political process is reflecting correct values. The success of the California State Water Bank in 1991 and again in 2009 is proof. When drought diminished water supplies within the state to critical levels, most of the water sold to the bank came from farmers who fallowed their land or replaced their supplies with groundwater or stored water. The water bank concept could also be used in other contexts such as addressing shortages within the Colorado River basin or accommodating fisheries in the Columbia River basin.

Burness and Quirk (1980, 133) asserted that "often what appears to be a shortage of water is actually the manifestation of restrictions on water rights transfer." Innovative water transfers are possible only if state water law and federal reclamation law become less restrictive. The past few decades have seen substantial progress toward streamlining water markets. Markets have been accepted as a good way to address water allocation problems resulting from drought and changing societal attitudes toward the environment. States have taken steps to streamline their transfer processes while protecting third parties.

If irrigators are permitted to sell their water, they would have the incentive to consider more water-efficient irrigation technologies and cropping patterns. Unfortunately, the advocates of increased governmental restrictions on transfers are driven by a market failure perspective that fails to take into account the flexibility of the prior appropriation system and the significance of incentives. Rather than giving water users the incentive to increase efficiency by minimizing restrictions on transfers and thereby freeing up water to be applied to newly valued uses, market opponents seek to regulate water away from current users under the guise of fairness to third parties without standing in the prior appropriation system. Forced reallocation of water, however, is more characteristic of the recently-fallen institutional systems of Eastern Europe and the former Soviet Union than it is of a market-based system.

This chapter has demonstrated that efficient water allocation may be achieved by salvaging the appropriation doctrine rather than imposing more and new institutions. Many water demanders are now using markets, rather than litigation, to achieve their water allocation objectives. Indeed, the prior appropriation doctrine has made a comeback (Huffman 1991). However, more freeing of the doctrine is needed to fully unlock the potential of voluntary water trading. The problem is that existing bureaucratic agencies lose power when allocation is turned over to the market process. To develop water markets, ideals must change and new political coalitions must reduce, rather than increase, the power of bureaucracies. When this happens, the doctrine of appropriation will provide the institutional foundation for gains from trade.

6

THE NEW FRONTIER

Historically, eastern states were blessed with a seemingly unbounded supply of water. Until recent decades, competitive demands among environmental, industrial, and residential water users generally could be met with available supplies and conflicts over water were "rare and short-lived" (Dellapenna 2002, 9). However, decades of rapid population growth, erratic precipitation, and prolonged drought have made clear the fact that water is indeed a finite resource in the humid East.

In 2007, parts of the southeast experienced the worst drought conditions in over one hundred years (Wang et al. 2010). By November, Lake Lanier, Atlanta's primary source for drinking water, held only enough water to supply residents for roughly 90 days. About 150 miles northwest, in Orme, Tennessee, water supplies did go dry requiring water to be trucked in from nearby Alabama (Glennon 2009, 95). Following the drought in North Carolina, crop losses prompted the U.S. Department of Agriculture to declare 59 counties as disaster areas (North Carolina Division of Water Resources 2008). With expectations of continued population growth and increasingly uncertain supplies, water shortages are unlikely to go away.

The imbalance between water supplies and demands has not only spawned water shortages, but also conflict among states and over various water uses. Since the 1980s, disputes between Alabama, Georgia, and Florida over access to water from the Apalachicola–Chattahoochee–Flint River Basin have intensified (Ruhl 2005). At the core of the conflict is growing competition for water among urban, industrial, and environmental demands. In Georgia, the Chattahoochee River provides drinking water to Atlanta's nearly 4 million residents and hydropower to 2.3 million customers before connecting with the Flint River, which supplies needed water for agricultural production in southern Georgia. As the Chattahoochee winds along Alabama's eastern border, it provides flows for hydropower and recreation. In Florida, the Chattahoochee and Flint rivers converge to form the Apalachicola, which empties into the Gulf of Mexico and provides critical ecological habitat for economically important and endangered species. Because no state has clear rights to a specified amount of water, disputes must be resolved through the courts. Not surprising, each state has filed suits against the other on behalf of each state's purported interest in water (Associated Press 2010).

Because scarcity-driven water shortages and conflicts have only recently proliferated in the eastern United States, achieving efficiency and conflict resolution in riparian states is more challenging and implies more drastic reforms than were suggested in Chapter 5.[1] The prior appropriation doctrine was born of desert and high prairie. Hence, water markets are more prevalent in the West than in the East. Where markets are active, water prices tend to more closely reflect opportunity costs and scarcity values. Not only do markets tend to encourage water use efficiency, they afford gains from trade and reduce the tendency for fighting in the courts and legislatures over who gets to use the water.

The geographic disparity in water markets is a product of water scarcity and the resulting institutions that evolved over time. As the West was settled, the need to move water away from streams led to water rights that could be separated from riparian lands. And to reduce conflict and provide for reallocation of water, those rights are more clearly defined and transferable than in the East. Though the early water rights tended to favor diversions for mining and agriculture, those rights evolved to include municipal and industrial uses and, more recently, instream flows for environmental and recreational uses. As with so many property rights, necessity was the mother of institutional invention.

In sharp contrast to the West's aridity, riparianism originated in water-rich England and took hold in the eastern United States during a period when water supplies dwarfed water demands. The riparian doctrine is thus premised on the assumption that water is and would always be plentiful. Prolonged drought and population growth in the East have proven the assumption of water availability incorrect and that riparianism is ill-equipped to efficiently allocate scarce water.

This chapter discusses the continuing evolution of the riparian doctrine in terms of how it will affect water-use efficiency and whether it will abate scarcity-driven water conflicts in riparian states. It begins with an explanation of how the riparian doctrine has evolved over time, from the natural flow theory to reasonable use standard to regulated riparian model. The chapter then describes why the emerging trend toward regulated riparianism is an improvement over strict traditional riparianism, but may still exacerbate water crises rather than resolve them. It concludes with a discussion of which institutional reforms are needed to salvage the riparian doctrine in the face of growing water scarcity.

From Natural Flow to Reasonable Use

For much of the eighteenth and early nineteenth centuries, water supplies in eastern rivers, lakes and aquifers dwarfed demands on those resources (Butler 1985, 96–97). Consequently, riparianism and the natural flow rule emerged as an appropriate and effective system to manage water. Though conflicts over water were not unknown, especially as mills became more numerous during the eighteenth century, they tended to be small in scale and short in duration (Caponera 2007). In general, it was possible for all riparian landowners to enjoy an undiminished quantity and quality from the natural flow of the river.

The efficacy of the natural flow standard can be understood in terms of what economists call the rivalrous nature of a good. Rivalrous goods are those for which one use precludes another. For example, person A's consumption of a loaf of bread precludes person B from consuming the same loaf. When goods are rivalrous in consumption, institutions (e.g., communal sharing, markets, or regulations) must determine who gets to use the good lest conflicts erupt.

In the West, aridity and diversions made water use rivalrous from the beginning. One person's diversion of water from a stream precluded or at least reduced another user's diversion. To prevent conflict between rivalrous uses, the West's prior appropriation doctrine allocated private water rights of defined quantity and priority.

In contrast to rivalrous diversions in the arid West, water use in the humid East tended to be non-rivalrous because water was more abundant and because its use by one riparian landowner—primarily domestic consumption or stock watering— did not preclude use by another person further downstream. Given the initial lack of rivalry, the natural flow doctrine worked well to meet the institutional needs of riparian owners.

As in the West, growing competition and new water demands became the mother of institutional reform in the East. In the nineteenth century, as industrial and municipal demands grew so too did the need to secure access to greater quantities of water, at times for non-riparian uses. The inherent nature of the natural flow rule proved to be an impediment to such development. It was no longer feasible to maintain an undiminished quantity and quality of water flowing in rivers and streams, and proved too strict as industrial and municipal diversions increased (Rose 1990).

But rather than developing a system of private, quantified, and prioritized rights as western courts had done with the prior appropriation doctrine, courts in eastern states replaced the rigid natural flow standard with the flexible reasonable use doctrine. This rule guaranteed each riparian owner the equal right to make reasonable uses of the watercourse (Butler 1985, 107). As explained by the Massachusetts Supreme Court, a "proprietor may make any reasonable use of the water of the stream in connection with his riparian estate and for lawful purposes within the watershed, provided he leaves the current diminished by no more than is reasonable, having regard for the like right to enjoy the common property by other riparian owners."[2] In short, the reasonable use doctrine softened the natural flow theory's strict prohibition on impacting stream flows so that more consumptive uses could occur.

Since adoption of the reasonable use standard, state courts have enumerated various factors for determining the reasonableness of a disputed use (Caponera 2007). The Restatement (Second) of Torts §850A summarizes the factors:

(a)   the purpose of the use;

(b)   the suitability of the use to the watercourse or lake;

(c)   the economic value of the use;

(d)   the social value of the use;

(e)   the extent and amount of harm it causes;

(f)   the practicality of avoiding the harm by adjusting the use or method of use of one proprietor or the other;

(g)   the practicality of adjusting the quantity of water used by each proprietor;

(h)   the protection of existing values of water uses, land, investments and enterprises; and

(i)   the justice of requiring the user causing harm to bear the loss.[3]

The number and subjectivity of these factors give courts more discretion in adjudicating water rights than was allowed under the natural flow theory.

Because courts consider "the reasonableness of a particular use not only in isolation, but also in comparison to other potentially reasonable uses of water" (Klein et al. 2009), water rights under the reasonable use doctrine are correlative and thus less clearly defined than quantified rights with a defined priority. During times of water shortage, courts administering the reasonable use doctrine attempt to maximize the social benefit of water use while minimizing the harm to other users.[4] Proportional sharing of withdrawal reductions is the preferred approach.[5] However, when water scarcity requires that some uses be entirely cut off, the economic value of the rivalrous uses appears to be the most important factor (Dellapenna 2004, 317). Therefore, courts rather than markets determine the highest and best use of scarce water resources.

This approach to allocating scarce water creates uncertainty for water users. As legal scholar Dellapenna (2004, 317) explains, with "vague and unpredictable criteria for decisions, even long established uses can be cut off without compensation if a court decides that a recently begun use is more reasonable." In addition, dispute resolution under the reasonable use standard is inherently unstable because the economic value of rivalrous uses can change over time (Ausness 1986, 416–18). According to Butler (1985, 126), the "reasonable use requirement is imprecise and varies according to the facts and circumstances of a particular situation. Because of this imprecision, and flexibility, users have had difficulty predicting whether a use will be protected in the future." The uncertainty and instability of riparian rights discourages investment in water infrastructure and water marketing, and it tends to favor large volume users with the financial resources to endure protracted litigation (Dellapenna 2004, 317).

Courts and legislatures in the eastern United States further reduced the clarity and security of riparian water rights by decoupling water use from riparian zones. Under a strict interpretation of the riparian doctrine, water can only be used on lands adjacent to the stream. To accommodate demands by non-adjacent users, however, every eastern state has recognized the right of municipalities to condemn water for public water supply purposes, regardless of whether the water is used on riparian land (Christman 1998). Though water use on non-riparian land was per se unreasonable under the original reasonable use doctrine (Dellapenna 2004, 316), the standard has changed to allow non-riparian[6] and intra-basin municipal uses.[7]

Some states have also recognized prescription and conveyance as legal means of procuring water rights for non-riparian uses. Prescriptive water rights are acquired in the same way that an adverse possession claim is established over real property—by using water continuously, openly, and in a manner hostile to the riparian landowners for a statutorily defined number of years (Dellapenna 1991, 308). Rights acquired by conveyance—via buy–sell agreement or lease with a riparian landowner—are recognized in only a handful of states[8] but represent another expansion of use rights in riparian jurisdictions. These expansions of the reasonable use doctrine not only increase the demands on eastern watercourses, they erode the exclusivity of use imposed by the appurtenancy requirement.[9]

In essence, the reasonable use standard and the dilution of the appurtenancy requirement have reduced what professor Henry Smith called the exclusionary nature of riparianism (Smith 2008). By exclusion Smith means that those with rights to water—whether those rights are based on prior appropriation or riparianism—can exclude non-owners from claiming and using the resource. By requiring riparian owners to prove that their uses are reasonable and by allowing non-riparian owners to claim water by prescription, conveyance, or on the grounds that their uses are more beneficial, the reasonable use standard substitutes a subjective governance system for more clearly delineated exclusionary rights (Smith 2008).

According to Smith, the transition from exclusion to governance may have been necessary—and even efficient—given the high costs of measuring, monitoring, and enforcing exclusionary rights to use water. However, as Smith explains, the natural flow theory "was easy to administer and may well have served as a baseline from which parties could contract in situations in which transactions costs were not prohibitive" (Smith 2008, 473). Though governance has allowed more consumption than might have resulted under a strict application of the natural flow standard, it has reduced the clarity and security of riparian water rights and, consequently, the potential for water trading in the East.

From Reasonable Use to Regulated Riparianism

As part of the transition from exclusion to governance described by Smith, several eastern states began adopting permit-based approaches to water allocation as early as the 1950s (Tarlock 2004). This approach, called "regulated riparianism," reflects the administrative nature of the allocation process (see Dellapenna 1991). It does not supplant the reasonable use doctrine, but rather creates a pre-conflict administrative process for determining who has the right to withdraw water from streams.

Though various eastern states have taken different approaches to regulated riparianism,[10] all regulated riparian systems share several fundamental deviations from the natural flow or reasonable use doctrines. The most significant of these are allocation of permits through administrative agencies, further relaxation of restrictions on place of use, and recognition of temporal priority in water use (Klein et al. 2009, 8). These departures from the strict riparian doctrine can encourage water markets and improve water-use efficiency, though, like all institutional reforms, the devil is in the details.

Consider first how the allocation of permits through an administrative system can encourage water marketing. When water-use rights are granted under reasonable use standards, they evolve as disputes occur and therefore are correlative— that is, conditioned by pre-existing use rights.[11] Regulated riparianism, on the other hand, specifies use rights before water disputes arise, so the rights are only correlative in the sense that one rivalrous use cannot harm another user's rights. To the extent that permits granted under regulated riparianism clearly specify use rights and disallow harm to third parties with rights to use water, the permitting process can lower transaction costs and encourage gains from trade (Abrams 1990).[12]

A second benefit of the permitting process concerns the allowance and clarification of water use outside riparian zones. Strict riparianism only allowed water use on lands adjacent to streams. This limitation was not a problem when the population was concentrated along watercourses but resulted in inefficiencies as populations expanded to non-riparian lands. As a result, states have diluted the appurtenancy requirement as they transitioned to regulated riparianism (Flood 1998). Several states even allow "interbasin transfers from one watershed to another—an idea that would have been anathema under the common law" (Klein et al. 2009, 10).[13] To the extent that abandoning the appurtenancy requirement allows for higher valued, non-appurtenant water uses, this element of regulated riparianism can increase water allocation efficiency.

A third way in which regulated riparianism helps clarify rights and promotes the potential for water markets is by implicitly recognizing temporal priority. Such recognition has a long-standing tradition in common law known as "first possession" (Lueck 1995; Eggertsson 2003). First possession is a low transaction cost way of assigning rights to those who are first to put a resource to beneficial use. More subtle than the "first in time, first in right" principle of the prior appropriation doctrine, recognition of temporal priority in regulated riparian states is manifest in the "first possession" requirement that new water users not interfere with existing water uses. For example, Florida law requires as a condition of permit issuance that applicants demonstrate that "the proposed use of water . . . [w]ill not interfere with any presently existing legal use of water."[14] By making new water permits subject to earlier granted permits, regulated riparianism can clarify priority of use.

The benefits of regulated riparianism for encouraging water markets are partially offset by two factors. First, dilution of the appurtenancy requirement may weaken eastern water institutions by allowing opportunistic claiming and by allowing more parties to object to permits on the grounds that they are harmed. The common law's appurtenancy requirement created an exclusive community of water users with relatively low information and coordinating costs (Smith 2000; Choe 2004).

These costs rose as exceptions to the reasonable use doctrine emerged, most notably when non-riparian users established rights through eminent domain, prescription, and conveyance. By further abandoning the appurtenancy requirement, eastern states implementing regulated riparianism expanded the universe of potential water claimants and the complexity of the interactions between them.[15]

A second downside to regulated riparianism is its potential for encouraging rent seeking, as explained in Chapter 4. Because the permitting process allocates resource rents associated with water use, it encourages people wanting to capture this value to expend time and money trying to convince the agency that their use is more valuable than competing ones. Though rights and the resource rents therefrom are ultimately allocated, the process can be expensive and there is no guarantee that the rights will go to the highest value users.[16]

The potential for rent seeking and agency capture is worsened by the broad discretion granted to administrate agencies in some jurisdictions. Whereas the common law standards of natural flow limits use to riparian landowners and whereas reasonable use weighs new use rights against their impact on existing rights, permitting processes generally consider a multitude of subjective public interest factors including economic, social, and environmental impacts, all of which are difficult to quantify (American Society of Civil Engineers 2004).

The type of subjectivity inherent in issuing permits takes us back to Henry Smith's (2008) emphasis on the importance of exclusion in any water rights system. Recall that exclusion means that those with rights to water—whether those rights are based on prior appropriation or riparianism—can exclude non-owners from claiming and using the resource. In the absence of exclusion, the ultimate form of rent seeking results from the tragedy of the commons.

As Smith argues, governance is a substitute for exclusionary rules, but it potentially creates another type of tragedy, namely what has been called the "tragedy of the anti-commons" (Heller 1998). The tragedy in this case is not that there are too many parties competing to use water, but rather that there are too many parties preventing its use. Imagine a case where 100 people own a parcel of land and each has the right to veto any change in use. Under such circumstances, we should not be surprised if land lies idle or is not put to its highest valued use.[17]

The tragedy of the anti-commons arises in water use if there are few limits on who can claim to be harmed by new water claims or changes in use. Exclusionary systems limit claims of harm to those with standing based on bona fide rights. For example, the prior appropriation system generally allows existing water rights owners to contest changes in use or claims for new rights if they believe they will be harmed. If harm is proved, new uses and new rights are disallowed; if not, they go forward. Though this adds transaction costs, such costs are a necessary part of clarifying property rights.

Accurately accounting for return flows is a case in point. Because return flows are used and reused along a river's course, altering points of diversion and quantity of consumptive use could harm downstream rights holders. Under the prior appropriation doctrine, the contested case hearing process allows potentially harmed parties to protest changes in use that could change return flows. Under the strict riparian doctrine, any change in return flows could be construed as a harm to downstream users.

As the riparian doctrine has evolved to accommodate non-riparian consumptive uses, the potential for third-party harm has become more likely as well as more complex. Allowing riparian users to transfer water to non-riparian lands in excess of the historical consumptive use obviously leaves less water available for downstream rights holders. The added costs of administering non-riparian water uses coupled with the higher probability of conflict between distant users could dissipate the efficiency gains of allowing non-riparian uses.

Expanding the set of potentially harmed parties beyond those with formally recognized rights can exacerbate the potential for a tragedy of the anti-commons. Under the prior appropriation doctrine this has occurred through the public trust doctrine as discussed in Chapter 5. In the case of regulated riparianism, it is happening as reasonable use considerations go beyond riparian landowners and correlative rights to include subjective public interest factors.

Regulated riparianism has the potential to increase the clarity, security, and transferability of water rights in the East, but the gains can be dissipated if the administrative allocation of water permits becomes mired in rent seeking. Similarly, if abandoning the place of use restrictions allows for water transfers that harm other rights holders, then litigation costs are likely to overwhelm allocation efficiency gained by allowing non-riparian uses. Finally, if the recognition of temporal priority fails to match permitted uses with physically available water, conflicts are likely to be more numerous and contentious. In short, regulated riparianism has the potential to dramatically increase the efficiency of water allocation in the East, but only if the proposed reforms result in more clearly defined, enforceable, and transferable water rights.

Taking the Plunge into Water Markets

As discussed earlier, some elements of regulated riparianism have clarified water use rights and increased the potential for using water markets in the East. The definition of individual permits provides more clarity than the common law's correlative right structure, the recognition of temporal priority has the potential to increase the security of those permits, and the dilution of place of use restrictions makes water more transferable. To a degree, these reforms have been enumerated in the Regulated Riparian Model Water Code, a comprehensive legislative scheme last published by the American Society of Civil Engineers in 2004. In particular, the Model Code proposes a detailed framework for allocating water permits and resolving disputes between water users. It also encourages water trades, including interbasin and interstate water transfers, and it explicitly rejects the riparian doctrine's strict appurtenancy requirement (American Society of Civil Engineers 2004, §1R-1-07 and §2R-1-02).

However, the Regulated Riparian Model Water Code falls short of laying the institutional foundation for water marketing in the East. For water markets to emerge as a way of mitigating conflicts over scarce water in the East, additional institutional reforms are necessary. Adding the following elements to water institutions in the East will help clarify water-use rights, reduce the cost of water market transactions, and increase the efficiency gains therefrom.

Grant Perpetual Permits and Treat Them as Legal Rights

As currently structured, permit-based riparianism in most states fails to define private property rights in water such that water users are likely to trade. The Regulated Riparian Model Water Code is demonstrative. Under the Model Code withdrawal permits are issued for a fixed term of 20 years and revocable upon an administrative finding that the use is unreasonable, i.e., that an alternative use of the water is more desirable from society's perspective (§7R-1-02). As the comment to §1R-1-05 explains, "water rights under this Code are not some form of private property which the State is debarred from interfering with without paying full compensation," but are instead permits of limited duration that are revocable during times of scarcity or when the use is determined to conflict with the public interest (§7R-3-01 and §1R-1-05–06).

Thus, while reasonableness remains the standard by which withdrawals are evaluated, the shift toward a permit system actually weakens existing riparian water "rights" by requiring users to regularly defend their water usage as being in the public interest (§6R-3-02).

Such requirements undermine the security of riparian rights and the likelihood of trading. The revocable nature of the withdrawal permits adds uncertainty to an already uncertain delineation of water rights; and by casting such doubt over the continued existence of the underlying asset, very little trading has occurred in regulated riparian states and very little trading can be expected.

Codify Transferability and Require Legal Standing for Challenges to Proposed Transfers

The permissibility of water transfers in regulated riparian states is far from certain. The natural flow theory of common law riparianism allowed for the possibility of private ordering in water allocation but permit-based riparianism grants exclusive allocation authority to government agencies by requiring agency approval of any permit modification or transfer (§7R-1-02). For instance, the Regulated Riparian Model Water Code states that transfers are permissible, but only when the proposed transfer is consistent with the public interest (§1R-1-07). Though agency approval might be a necessary and low-cost means of protecting other rights holders, the ability of users to transfer rights and the criteria for determining transferable volumes must be made explicit.

Furthermore, under the Model Code's requirement that agencies consider the societal value of a permit transfer and its expected impact on the public interest, any water trading that occurs in regulated riparian states must garner the approval of not only the administrative agency, but society as a whole (§7R-1-02, referencing

§6R-3-02). This creates an anti-commons out of water resources in regulated riparian states in which exclusion or veto rights are too numerous to allow for trading (Bretsen and Hill 2009). Challenges to proposed transfers should be limited to parties able to establish legal standing, that is, a cognizable connection to and harm from the transfer being challenged.

Prioritize Water Use According to an Objective Standard such as Temporal Priority

For scarce water rights to be clearly defined, legislatures, courts, or agencies must explicitly articulate which rights will trump others if available supplies fail to meet the quantity demanded. Temporal priority has been applied in the permitting process and is a potential standard for allocating water when permitted withdrawals exceed the volume of water available. When scarcity requires a reduction in water use, temporal priority provides greater certainty as to which users will be cut off and which can legally continue. Prioritizing uses by this standard reduces the uncertainty and subjectivity inherent in prioritizing by such considerations as social value and economic benefit. Of course, establishing a temporal priority for all water uses in riparian states could be very costly, but the resulting increase in clarity of rights would allow users to evaluate the security of their water access and hedge against future shortage.

An alternative to prioritizing rights temporally or by preferential use is to apportion reductions equally among all uses whenever scarcity requires it. Australia's water law allows for this approach by defining both water access entitlements and water allocations. An entitlement is a perpetual or ongoing right to a share of water from a specified consumptive pool, whereas an allocation is the specific volume of water supplied to water access entitlements in a given water year.[18] During times of shortage, administrative agencies reduce allocations to reflect the volume available for consumption. Different classes of entitlements vary in terms of reliability or the frequency with which the allocation equals the full entitlement, with a statement of reliability issued with each entitlement (Waye and Son 2010). The resulting clarity of rights and reliability provides water users with the flexibility to respond to shortage and seasonal risk either by reducing consumption or acquiring additional allocations (Brooks and Harris 2008).

Undertake Interstate Compacts

Because water knows no political boundaries, competition across political jurisdictions, especially state lines, can generate conflict. The great western explorer John Wesley Powell recognized this and suggested that political jurisdictions should be organized around watersheds, rather than fitting water institutions into political jurisdictions (Huffman 1997). Unfortunately, few municipal, county, or state boundaries follow Powell's suggestion, making it necessary to find ways to allocate water rights among jurisdictions. In the United States, this is accomplished through interstate compacting, a process whereby states sharing a major watershed agree on a division of water among the states with approval and enforcement undertaken by the federal government.

The Colorado River Compact is the earliest example of this process. This 1922 compact splits the Colorado River Basin into two areas and requires that the Upper Basin states (Wyoming, Colorado, Utah, and New Mexico) deliver at least 75 million acre-feet of water to the Lower Basin states (Arizona, Nevada, and California) every 10 years.[19] The compact further allocates the water to individual states as defined shares of the total flow. As explained in Chapter 1, this division of water resources at the state level is a prerequisite to the free market allocation approach proposed by the Colorado River Mitigation Bank. Without clearly defined water rights, states, like individual water users, have little incentive to conserve water or to maximize its efficiency by trading from low to high valued uses.

The interstate compacting process was employed when the states of Illinois, Indiana, Michigan, Minnesota, New York, Ohio, Pennsylvania, and Wisconsin signed the Great Lakes Basin Compact in an attempt to reduce competition for water in the Great Lakes Basin. Unlike the Colorado River Compact, however, the Great Lakes Compact does not apportion use rights among the eight signatory states. Instead, it prohibits, with limited exceptions, any diversion of Great Lakes water to points outside the Great Lakes Basin and requires states to implement water conservation programs.

This approach creates several major stumbling blocks for water markets. First, it does not provide mechanisms or even guidance for creating or clarifying individual water rights within states. Second, the compact makes it difficult to exchange water rights within the basin. Finally and most importantly, it disallows marketing water out of the basin. For these reasons, the Great Lakes Compact, which was signed into federal law in 2008, represents a missed opportunity for the Great Lakes states to clarify the interstate allocation of water and develop the institutional framework for water marketing.

To clarify water rights and encourage water markets, more eastern states will have to engage in the compacting process. Doing so is the only way to divide water among states within a basin, thus allowing individual states to establish secure water-use rights within their boundaries. Especially because municipal and industrial demands can straddle political boundaries, it will be important to avoid the pitfalls of the Great Lakes Compact that thwart water markets. Additionally, because waterways in the East provide important water transportation routes, it is important that future eastern state water compacts specify minimum flows for instream purposes.[20] Engaging in the compacting process before interstate conflicts worsen can help eliminate what legal scholar Joseph Dellapenna characterizes as "an unseemly hodgepodge of requirements" (Dellapenna 2002, 445) and help encourage both interbasin and interstate water trading by clarifying water rights and reducing transaction costs.

Conclusion

Increasing water scarcity has driven and will continue to drive institutional change. In the American West, this evolutionary process resulted in the prior appropriation doctrine. Though this evolution started more than a century ago, it allowed for better defined water rights and paved the way for water markets that are resolving water conflicts today.

Water abundance in the East staved off this evolutionary process because the riparian doctrine, steeped in English common law, worked well initially to accommodate non-consumptive water uses. All along a water course, people could harness water power, withdraw small quantities for domestic uses, and take advantage of low-cost water transportation. However, growing non-consumptive demands, as well as increasing consumptive withdrawals, are putting pressure on the allocation system.

The question is whether the evolution of water institutions in the East will tap the efficiency of water markets or get trapped in a quagmire of rent seeking. The current trend is toward water marketing as institutions have moved from a strict riparian doctrine to regulated riparianism. Because the former allowed little room for consumptive uses and no room for water transfers away from riparian zones, it had no way of promoting water marketing. Evolutionary pressures leading to regulated riparianism are opening the gates for water markets by establishing permits that quantify water-use rights and allow transfers to non-appurtenant regions. As with any process for establishing rights, there is always the possibility of rent seeking and always the possibility that regulation (governance) will thwart market exchanges. The potential efficiency gains as scarcity grows, coupled with the ability of water markets to encourage cooperation over conflict, however, will likely lead to more marketing, especially if the reforms suggested above are implemented.

7

BUY THAT FISH A DRINK

The John Day River Basin in eastern Oregon has a long history of mining, agriculture, and ranching. The river and its tributaries also provide habitat for spring Chinook salmon and summer steelhead, as well as the state's only population of westslope cutthroat trout. Persistent drought during the 1990s reduced stream flows, degraded water quality, and threatened the viability of local fish populations. The situation had all the makings of water conflict of the type witnessed in Oregon's Klamath Basin and California's Sacramento–San Joaquin Delta, where appropriative rights clash with public trust and endangered species claims.

The John Day, however, has become an example of how water markets, and particularly contracts for instream flows, can avert conflict and engender cooperation between competing water users. The story begins with the Oregon Water Trust (OWT), a private non-profit organization dedicated to restoring and preserving freshwater. In 2000, OWT entered into the first of five annual agreements to lease water rights from Pat and Hedy Voigt, third-generation ranchers along the John Day's Middle Fork. Per the lease agreement, the Voigts reduced their annual water diversions, leaving the water instream to improve fish habitat. The leases proved to be mutually beneficial, satisfying OWT's demand for improved flows while generating additional revenue for the Voigts (Columbia Basin Water Transactions Program 2006). In 2006, following permanent improvements in their water-use efficiency, the Voigts sold a portion of their water rights to OWT, protecting in perpetuity roughly 10 cubic feet per second (cfs) of water for fish.

Despite success stories such as the Voigts', the market for instream flows remains thin throughout the West. This is largely due to the region's institutional inertia favoring diversion-based water uses and the erroneous belief that collective action is necessary to protect stream resources (Anderson and Johnson 1983). As this chapter explains, markets can protect and even enhance stream resources when the legal institutions allow them to function. The chapter begins by describing the most significant institutional barriers to the private provision of instream flows. It then describes how several conservation groups have overcome these barriers and used private contracting to achieve restoration objectives where public agencies could not. The chapter concludes by outlining the different approaches western states[1] have taken to protecting stream flows, including recent legal reforms aimed at enhancing private transactions.

More Restraint than Failure

Two factors have limited private contracting for instream flows in the West. First, the prior appropriation doctrine evolved in a way that does not accommodate instream water uses. In particular, the doctrine's diversion and beneficial-use requirements preclude the definition, enforcement, and transferability of instream rights. The second factor relates to the misperception that instream flows are public goods which can only be provided through collective action. For instance, in many states, only public agencies can acquire and hold instream rights, a restriction that effectively crowds out private actors. These factors suggest that markets have not failed to protect instream flows, but rather that they have not been allowed the opportunity.

Institutional Inertia

According to legal scholar James Huffman (1983, 268), markets have played a relatively insignificant role in the provision of environmental flows not because they have failed, but rather because of deficiencies in states' water rights systems. Historically, private opportunities to protect stream flows were precluded by a legal system that evolved in accordance with the need to divert water from streams. The concept of beneficial use, initially developed for agricultural, mining, and domestic uses, did not recognize the numerous environmental, economic, and even social benefits to free-flowing streams. For example, in a 1917 case involving disputed ownership of instream flows for the purpose of supporting a duck population, the Utah Supreme Court held

> It is utterly inconceivable that a valid appropriation of water can be made under the laws of this state, when the beneficial use of which, after the appropriation is made, will belong equally to every human being who seeks to enjoy it. . . . [W]e are decidedly of the opinion that the beneficial use contemplated in making the appropriation must be one that inures to the exclusive benefit of the appropriator and subject to his dominion and control.[2]

This language suggested that the state was unwilling to allow individuals or groups to appropriate rights for the "public good."

The requirement that water had to be diverted to perfect a water right also interfered with the potential for using markets. For example, when the Colorado legislature authorized the Colorado River Conservation District to reserve water for instream purposes in any natural stream large enough to support a fish population, the Colorado Supreme Court ruled that there was "no support in the law of this state for the proposition that a minimum flow of water may be 'appropriated' in a natural stream for piscatorial purposes without diversion of any portion of the water 'appropriated' from the natural course of the stream."[3]

When the West was being settled, the diversion requirement served the purpose of notifying users on a watercourse of the amount of water controlled by an appropriator, as well as the location of the claimed water right. Diversion requirements also prevented speculators from claiming water for future sale. When states established record systems for water rights, the diversion requirement became obsolete but was nevertheless retained as an element of many states' appropriative water law systems. Unfortunately, as long as water use required a diversion, there was no way that property rights could be established for instream purposes.

Another facet of the prior appropriation system that can prevent the establishment of instream flow rights is the use-it-or-lose-it rule. In most states, water rights may be deemed abandoned or forfeited if the water is not "used," that is, if water is left in a stream. The law of forfeiture and abandonment is related to the beneficial-use standard in that it is designed to ensure that appropriated water is reasonably used and not wasted. Like the diversion requirement, it is also aimed at preventing speculation in water. For example, if an individual were to appropriate water from a stream in the hope that its value would rise, the water would not be used (diverted or consumed). The argument against speculation in water is that it causes valuable resources to remain idle and unproductive, thus inhibiting economic growth. This argument made sense when the West was being settled—resources were abundant and economic development was the common goal. The same argument makes little sense, today, however, because the individual appropriating a water right bears the cost of not consuming the water; his actions imply that he believes the water is more valuable left instream than diverted. By threatening water rights holders with loss of their rights if they do not consume the water, the law of forfeiture and abandonment stifles the establishment of instream water rights and discourages valuable instream uses.

Public Goods Problem

A second factor limiting the private provision of instream flows relates to the misperception that instream flows are public goods only capable of public provision. Public goods are defined as being (1) non-exclusive, meaning nonpaying users cannot be excluded from consuming the good, and (2) non-rivalrous, meaning once the good is provided to one individual, it can be provided to others at no additional cost. On the assumption that private parties will underproduce instream flows because of their non-exclusive and non-rivalrous nature, collective action becomes the default institutional arrangement. But if this assumption is wrong or if instream flows are not true public goods, then the case can be made for rejecting the collective action approach in favor of an instream flow market.

Consider first the non-exclusive nature of public goods. If users cannot be excluded from consuming the good, private firms will have no incentive to produce it because they will incur costs but receive little or no revenue from production. The ability to exclude nonpaying consumers depends on how much the supplier is willing to spend to keep people out. A rancher who owns a large section of land along a fishing stream may find it costly to deny fishermen access if they do not pay. Fences, signs, and game wardens provide some measure of enforcement, but they also cost money. Whether they cost too much and prevent private provision of fishing habitat depends on many variables including the technology of enforcement and the value of the resource. If the costs of exclusion decline or the value of the fishing experience rises, riparian landowners will invest more in exclusion and will be more likely to supply the fishing habitat, including the necessary instream flows.

The second characteristic of a public good is that once it is supplied to one individual, it can be supplied to others at no additional cost. For instance, stream flows can generate non-rivalrous values such as aesthetic values enjoyed by passers-by who prefer flowing to dewatered streams. Providing the view of a beautiful river to one individual does not preclude another from enjoying it at the same time, unless there is a crowding problem at some vista. Economic efficiency dictates that the price of the good should equal the marginal or additional cost of producing it. If the additional cost of providing the view for an extra person is zero, the implication is that a zero price should be charged. If so, a private producer is presumably unable to achieve a rate of return necessary to stimulate production.[4]

Markets on the Rise

Instream water uses may have public good characteristics and may be incongruent with the prior appropriation doctrine, but it does not follow that private markets cannot provide those uses. Specifically, the public good problem can be overcome by innovative contractual arrangements. Conservation groups such as Trout Unlimited, Ducks Unlimited, the Nature Conservancy, and private water trusts have demonstrated that private resources can be devoted to the provision of instream flows.

During the winter of 1989, for example, the Nature Conservancy (TNC) and the Trumpeter Swan Society acted to enhance instream flows for the benefit of wildlife on the Henry's Fork of the Snake River in Idaho. The resident population of trumpeter swans on the river was near starvation, its aquatic food supply cut off by river ice. Additional water from an upstream dam was desperately needed to open a channel and allow the birds access to their food. With voluntary donations, the conservation groups acquired releases of water from the reservoir valued at $20,000 to $40,000 and used them to enhance the flows in the river and save the swans (Anderson and Leal 1991, 108). Those who donated to the effort could not expect to exclude non-contributors from deriving an aesthetic or existence-based benefit, but they donated nonetheless and saved the swan population.

In another example, the Montana Water Trust entered into a 10-year lease agreement with irrigators in 2005 to reduce diversions along Tin Cup Creek in western Montana. The upper portion of Tin Cup lies within the Selway–Bitterroot Wilderness and provides critical native fish habitat, fostering westslope cutthroat and bull trout. The lower portion is heavily appropriated by irrigators and their diversions were depleting stream flows to levels insufficient for fish. After negotiating mutually beneficial lease agreements with several irrigators, the Montana Water Trust was able to acquire enough instream flows to reconnect migration routes between Upper Tin Cup Creek and the Bitterroot River downstream.

As these examples demonstrate, market processes can foster solutions to the public good problem. To the extent that instream flows seem non-rivalrous and non-exclusive, entrepreneurs have devised ways of collecting payment from beneficiaries and using them to achieve stream restoration objectives. Environmental entrepreneurs in organizations such as TNC, Trout Unlimited, Environmental Defense Fund, and various water trusts play an important role in creating private rights and capturing the benefits of the environmental amenities of instream flows. However, the ability of these conservation organizations to contract for stream restoration is ultimately limited by the definition of water rights under state law.

Beneficial-use standards, restrictions on who may acquire instream use rights, and laws that prohibit the transfer of conserved water continue to discourage market transactions. As long as rights to instream flows cannot be defined and enforced, markets cannot provide instream uses. Eliminating the obsolete elements of the prior appropriation doctrine discussed here and in Chapter 5 would lower the institutional barriers to the private provision of instream flows. Though governmental provision of instream flows remains the norm, the next section describes how several states have undertaken legal reform to unlock the potential of instream flow markets.

Instream Flows in Western States

Because most western states have declared in their constitutions or by statute that all waters within the state belong to the state and grant only usufructuary rights to water, and because instream uses have public good characteristics, it is not surprising that the standard approach to the maintenance of instream flows has been collective action. Rather than allowing private entities to acquire, hold, and transfer rights to instream flows, states have more commonly chosen to maintain instream flows by reserving water from appropriation, establishing minimum stream flows by bureaucratic fiat, conditioning new water permits, or directing state agencies to acquire and hold instream flow rights.

Fortunately, the laws in western states governing the use and allocation of water for instream purposes are evolving to better reflect new demands and remove the barriers to markets and private contracting. For instance, all western states now define certain uses of water instream as beneficial[5] and, in most states, water may be appropriated to an instream use or transferred from existing offstream right to an instream use. Despite these new provisions, states still rely heavily on central control of instream flows, often limiting acquisition and possession of instream rights to state agencies. However, this too is changing as states begin to rely less on restrictions and regulations that limit new water developments, in favor of market-based strategies and greater private involvement.

Arizona

In 1941, the Arizona legislature broadened the definition of beneficial use to include "wildlife, including fish" and, in 1962, added "recreation." Nonetheless, it was not clear until a 1976 court ruling that water could be appropriated for those instream uses.[6] Today, Arizona is one of only two western states (Nevada being the other) that permits private entities to appropriate water instream. The first permit was issued in 1979, when the Arizona Department of Water Resources approved TNC's application to keep water instream along Ramsey and O'Donnell Creeks (The Nature Conservancy 2006). Since then, of the 101 new applications filed with the state, 32 have been approved and the other 69 are still pending (Arizona Department of Water Resources 2009).

Because much of the state's waters are currently appropriated to other uses, new instream water rights are of limited use and effectiveness in preserving environmental or recreational amenities. By its very nature, a new instream appropriation can at best maintain existing flows, not improve them or restore flows to dewatered streams. Thus, new appropriations are most effective in situations where there is currently a sufficient amount of water not claimed by other users to support specific preservation objectives. On the other hand, market trades provide a means of improving or restoring stream flows. However, only the "state or its political subdivisions" can transfer existing offstream water rights to instream uses,[7] in effect prohibiting private interests in Arizona from trading water for uses other than for offstream applications.

In 1994, the state established the Arizona Water Protection Fund, a state grant program that provides funds for the "development and implementation of measures to protect water of sufficient quality and quantity to maintain, enhance and restore rivers and streams and associated riparian habitats, including fish and wildlife resources."[8] Funding for projects comes primarily from state appropriations but may include private donations and/or fees from interstate water trades. As of 2008, the fund had provided over $38 million for various restoration projects, benefiting more than 1,400 miles of rivers and streams throughout the state (Arizona Water Protection Fund Commission 2008).

California

California has adopted a mix of regulatory and market-based mechanisms to maintain or improve stream flows. To protect existing flows, the state relies heavily on regulatory restrictions governing new appropriations and transfers. The

State Water Resources Control Board (SWRCB) is authorized to deny or condition applications for new diversions. It also retains jurisdiction over water rights and may modify them to protect instream flows. In addition, changes of use that include water transfers are subject to instream flow protection conditions imposed by the board. Other indirect means available to the SWRCB include water quality requirements and reasonable-use requirements (Gray 1993). California's regulatory strategies have been criticized as inadequate for protecting instream flows because they essentially depend upon a case-by-case review of water right applications, which in turn depend upon the discretion of the SWRCB. Changing times and personnel on the board inevitably lead to inconsistency.

Amendments to the state's water code in 1991 and 1999 permit any person to acquire and change an offstream use right to instream for "purposes of preserving or enhancing wetlands habitat, fish and wildlife resources, or recreation."[9] Transfers may be permanent, temporary, or what are referred to as "urgency changes."[10] The transfer provisions provide opportunities to state, federal, and private entities for short- and long-term lease agreements and water right sales. Acquisitions of existing diversionary rights provide a means of improving stream flows to restore fisheries or habitat that may have been lost through historic offstream water uses.

Despite these market opportunities for non-governmental allocation of instream flows, few private transactions have occurred. Because of the complexity of the state's water laws and its administratively cumbersome transfer process, transaction costs can be prohibitively high for smaller trades (Boyd 2003). To date, market activity has been primarily limited to large-scale state and federal acquisitions.

In response to severe drought conditions which nearly pushed populations of delta smelt and winter-run Chinook salmon to the brink of extinction, the state created the CALFED Bay–Delta Program in 2000. It is a collaborative effort among 25 state and federal agencies with a common goal of protecting the ecological health of the Bay–Delta estuary, while maintaining water quality and reliable supplies to local farms and cities. As part of the program, the Environmental Water Account (EWA) was established to allow fishery agencies to augment stream flows for the protection of fish. Agencies acquire water from willing sellers and then alter water releases from state and federal water projects to provide adequate flows for target fish species. In 2007, the EWA acquired 275,000 acre-feet from willing sellers and an additional 215,000 from state reservoirs to increase flows for delta smelt and Chinook salmon (California Bay–Delta Authority 2008).

Colorado

In Colorado, the Colorado Water Conservation Board (CWCB) is authorized to apply for unappropriated rights to maintain instream flows or to acquire existing rights by "grant, purchase, donation, bequest, devise, lease, exchange, or other contractual agreement, from or with any person, including any governmental entity."[11] The Board is specifically prohibited from acquiring rights by eminent domain. The Colorado system has two important characteristics. First, like any other appropriator, the Board must have evidence that its requests for water constitute a beneficial use. Second, when the Board purchases existing rights, it must pay the market rate. Though the state's budget constraint differs considerably from that of private users, the system does force the state to consider the costs of instream flows. In a state such as Colorado where most water rights have already been claimed, the market approach is probably the only viable alternative to state action. However, Colorado falls short of creating an extensive instream water market because only the CWCB is permitted to hold instream flow rights.

Despite the state's limitations on who may hold instream flow rights, private innovation has proved fruitful. Private conservancies such as Trout Unlimited (TU), the Colorado Water Trust (CWT), and TNC have played an active role in preserving and restoring habitat and flows for fish and wildlife species. In 1992, TNC was given 300 cfs in senior water rights that were then donated to the state for instream use, protecting flows along 29 miles of the Gunnison River, including a section of the Black Canyon of the Gunnison National Park (Colorado Water Conservation Board 2008). The CWT acquires water rights from offstream users along ecologically important stream channels and then donates them to the CWCB to be transferred to instream use. In addition, they work closely with existing water users to implement water conservation measures that preserve current diversionary uses while improving wetlands, fish and wildlife habitat, and open space amenities. TU's Colorado Water Project has been working with state policymakers, discussing the possibility of new legislation that would permit private entities to temporarily lease water for instream use without having to donate the rights to the state (Colorado Trout Unlimited 2008).

Idaho

In Idaho, protection of instream flows comes primarily from the state. In 1978, the state legislature passed a law that established base flows on the Snake River and authorized the Idaho Water Resources Board (IWRB) to appropriate water for instream uses. The law authorized the IWRB either to initiate action or to respond to private requests for instream appropriations. While actions of the IWRB are subject to technical constraints such as past flow records and minimum flows necessary for the preservation of fish and wildlife habitat, recreation, navigation, aesthetic beauty, and water quality, the board still has a great deal of discretion to act in the public interest. "As of February 2010, there were 297 licensed or permitted water rights for minimum stream flows covering about 1577 miles of streams and four minimum lake levels" (Idaho Department of Water Resources 2010a). This is less than 1 percent of the total stream miles in Idaho.

The board has been more successful at protecting instream flows under the 1988 Idaho Protected Rivers Act that calls for designation of specific river reaches as either natural or recreational. These designations effectively create minimum stream flows by prohibiting construction, diversion, and alteration activities on the watercourse. The studies used to determine whether to designate rivers as protected are also used by the IWRB as a planning vehicle in seeking instream flow permits (Beeman 1993, 13–15). Since 1988, over 2,745 miles of streams have been protected under the Act (Idaho Department of Water Resources 2010b).

In addition to state protection of minimum flows and instream appropriations, state and federal agencies lease and purchase water rights from willing sellers to restore flows for endangered or threatened fish species. The water is generally acquired through local rental pools where farmers can lease water on a temporary basis to the state or federal government. The U.S. Bureau of Reclamation is the most active acquirer of water (up to 487,000 acre-feet per year) to augment flows for salmon and meet requirements of the Endangered Species Act.

Market opportunities for private entities are limited, though improving. Legislation passed in 2007 allows private right holders to donate all or a portion of water rights to the state to be held in trust for the preservation of minimum flows along the Big and Little Wood Rivers.[12] Other options to protect flows require some innovation. For example, Trout Unlimited (TU) provides funding and assistance to landowners for projects that improve stream flows indirectly. By improving water use efficiencies, altering timing and location of diversions, and implementing conservation practices, landowners are able to maintain and even improve productivity, while decreasing the amount of water pulled from streams. In 2007, TU completed a 3-year restoration effort with local landowners that included changing a diversion point and providing funding to several ranchers to switch from flood to sprinkler irrigation. The decrease in diversions restored flows to Badger Creek and reconnected 6.4 miles of critical spawning habitat without reducing agricultural production (Trout Unlimited 2008).

Montana

Montana has maintained instream flows primarily by reserving water from private appropriation. The 1973 Water Use Act authorized federal, state, and local governments to apply to the Department of Natural Resources and Conservation (DNRC)[13] for water reservations for existing or future beneficial uses, including the maintenance of minimum stream flows. The DNRC's most ambitious undertaking was its consideration of applications for reservations on the Yellowstone River in a single proceeding. According to Huffman (1983, 264),

> Because private water users could not apply for reservations, the board sought to assure that the reservations that it granted did not tie up all of the water and thus prohibit any future private development of water. However, the variable nature of the stream flow and the inadequacy of much of the data available to the board raises some doubt about the prospects for future private water development.

Though the department granted numerous reservations for consumptive water uses by municipalities and irrigation districts, by far the lion's share of Yellowstone

River water was reserved for instream flows. The instream flow reservations were granted on approximately 2,078 stream miles in the Yellowstone River Basin. These flows constitute 60–70 percent of the average annual flow in the basin and approximately 12.5 percent of Montana's total stream miles (McKinney 1993, 15–16). Currently, the Montana Department of Fish, Wildlife and Parks (FWP) maintains minimum flow reservations on various reaches of 87 streams (Amman 2008).

Because the reservation system allows little room for economic criteria and little flexibility once the reservations are made,[14] market mechanisms have been introduced into the law. In 1989, the Montana Legislature granted FWP authority to lease water rights to maintain fish flows on a limited number of streams. Legislation passed in 1991 and renewed in 1999 expanded FWP's authority to streams throughout the state. Authorization for the lease may only be granted after notice of the application is published and all objections resolved. The authorization procedures are essentially the same as those for any change of an appropriative water right.

In 1995, legislation created a 10-year pilot project to extend the option to lease water for instream flows to the private sector. Despite limitations on leasing terms (maximum of 10 years) and some uncertainty about the future of the program, private conservation groups such as TU and the Montana Water Trust took full advantage of the new marketing opportunities to improve stream flows for fish.

The Montana Water Trust embraces a free market approach to restoring dewatered fisheries throughout Montana, acquiring water rights through short- and long-term contracts and entering into cooperative agreements with landowners that improve water-use efficiencies and flows for target fish species. Through the use of remote stream flow monitoring technologies, the trust ensures flows targets are maintained while reducing monitoring and enforcement costs.

Montana TU and TU's Montana Water Project have been involved in nearly every aspect of conservation and restoration of the state's fisheries, from education and on-the-ground restoration projects to water right acquisitions and drafting legislation to improve habitat and stream flows throughout the state. TU's portfolio of long-term instream leases along tributaries to the Blackfoot River restores flows and improves habitat for native fish species.

With the success of the program and support from a coalition of environmental and agricultural interests, the legislature, in 2005, made permanent the option for private and public entities to lease water for instream purposes. Montana's evolving laws are a testament to the viability of private provision of instream flows.

Nevada

Nevada has taken a unique approach to instream flows and marketing. Unlike other western states, opportunities to protect stream flows in Nevada stem from common law. A 1988 State Supreme Court's decision held instream uses of water for fish and recreation were considered beneficial. Together with state water statutes, it has provided for temporary and permanent acquisitions of water for instream use and anyone may apply to appropriate or transfer water for instream purposes. The transfers may be temporary or permanent, providing leasing or outright purchases of water rights to restore and protect stream flows.

TNC, in cooperation with the state, U.S. Fish and Wildlife Service, and the Nevada Waterfowl Association, have purchased nearly 30,000 acre-feet of water rights from farmers in the federal Newlands Project to leave instream in the Carson River. The beneficiary of these instream flows is the Stillwater National Wildlife Refuge into which the Carson River empties. At one time, the refuge was a huge marsh covering an average of 100,000 acres in the Lahontan Valley at the base of the Sierra Nevada Mountains and was a major stopover for hundreds of thousands of migrating birds. Irrigation diversions by the project gradually reduced it to 3,100 acres in 1990. The goal is to restore 25,000 acres to productive wetlands. TNC's water purchases were made possible by the federal Truckee–Carson Water Rights Settlement Act, which instituted a program of voluntary purchases of water rights and land with water rights from willing sellers. TNC's response to the plight of the Stillwater Refuge clearly demonstrates that the demand for instream flows to preserve wetlands and protect wildlife can be met by private entities.

New Mexico

Instream protection of water in New Mexico comes primarily as a by-product of interstate compacts and directly from efforts to sustain endangered species. To fulfill its obligations under the Pecos River Compact, the state's Interstate Stream Commission leases and purchases water rights from primarily agricultural users and leaves the water instream to flow into Texas. Although not intentional, the acquisitions improve instream flows for fish and recreational uses along stretches of the Pecos River.

State and federal agencies may also acquire water from willing sellers to improve flows for the protection of endangered species. The U.S. Bureau of Reclamation has leased water from the state (water often acquired from willing sellers) to improve flows for the silvery minnow. Legislation adopted in 2005 created and funded the Strategic Water Reserve, authorizing the state's Interstate Stream Commission to lease, purchase, or receive as a donation water rights to help improve and protect instream flows for endangered species, in addition to meeting interstate stream compacts.

Opportunities for private demanders of instream flows are less clear in New Mexico. The state's waters are fully appropriated to other uses, leaving transfers of existing rights as the most viable means to protect instream flows. Although the state's legislature does not explicitly recognize water instream as beneficial, the attorney general issued an opinion in 1998 concluding that the state engineer can protect instream flows for "recreational, fish or wildlife, or ecological purposes."[15] The opinion created the possibility for transfers of current consumptive water rights to instream uses. However, decisions by the state engineer are subject to review by the courts which may ignore the attorney general's opinion (Boyd 2003). Legal ambiguities and long-term uncertainty over whether flows can be protected remain as disincentives to private entities stepping forward.

Oregon

Oregon first passed instream flow legislation in 1955 that allowed the state, through an administrative process, to set minimum flows to protect salmon during spawning season. Water designated as minimum flows was protected instream and unavailable for appropriation to other uses. By 1987, the state had set more than 500 minimum flows on various streams; however, despite the state's efforts, the act fell short of providing reliable protection of flows. Because minimum flows protected only unappropriated water they were inherently incapable of restoring dewatered streams or improving flows where needed. Because much of the state's waters were already appropriated to offstream users by the 1950s, minimum flows, at best, protected what little water remained instream. Moreover, minimum flows were junior to other water users, thus in years of low flows there was no guarantee target flows could be maintained.

Then in 1987, the state passed a comprehensive Instream Water Rights Act, which opened the door for new instream flow rights and market trading of water for environmental uses. The law provides greater legal protection of water instream and opportunities to restore previously dewatered streams. The state through the departments of Fish and Wildlife, Environmental Quality, or Parks and Recreation, are authorized to apply for new instream water rights. The act also directed the Water Resources Department to convert all minimum perennial streamflows that were established under the 1955 Act to instream flow rights. The new water rights are similar to minimum flows, though despite their limited applicability and efficacy, they provide greater legal protection and clarity. As of 2008, more than 500 of the state's minimum flows were converted and over 900 state-applied instream rights have been appropriated (Oregon Water Resources Department 2008).

The law also provides several ways to improve instream flows through voluntary trades, including purchasing, leasing, or donating existing water rights for instream flows.[16] Any legal entity may acquire water rights from existing right holders, convert the rights to instream use and leave the water to flow instream. Once the acquired rights are changed to instream use, however, the rights are held in trust by the state. Because changed rights maintain the original priority date, they create enforceable senior water rights that reliably improve and protect stream flows.

The state's instream flow statutes have been a boon to private conservation groups. For the first time, private demands for instream flows could compete directly with traditional water demands in a market environment. Just as municipalities or industries acquire water from farmers, environmentalists or recreationists can lease or purchase water rights and leave the water instream.

The Oregon Water Trust (OWT), founded in 1993, has been a pioneer in market-based approaches to restoring stream flows. It actively acquires water rights using various leasing and purchasing agreements to restore flows for important fisheries throughout the state. In addition, OWT enters into innovative agreements with landowners that improve stream flows for fish, while maintaining agricultural production. Changes in the timing or location of diversions can reconnect stream flows through historically dewatered sections without significantly affecting total diversions for offstream uses. Through improvements in technologies such as switching from flood irrigation to sprinklers, production levels may be maintained or even improved while providing additional water instream for fish.

Other private entities such as the Deschutes River Conservancy (DRC) and Klamath Basin Rangeland Trust are working to restore flows and improve water quality in central Oregon and the Upper Klamath Basin, respectively. Both are dedicated to enhancing riparian ecosystem health through voluntary agreements, including water leasing, purchasing, banking, and cooperative efforts with landowners that restore flows without drying up agricultural communities. Between 1996 and 2008, the DRC restored more than 160 cfs to local streams and tributaries in the Deschutes Basin—"the equivalent of six Olympic-size swimming pools pouring into these rivers every hour" (Deschutes River Conservancy 2008).

Utah

Utah first passed instream flow legislation in 1986, authorizing the Utah Division of Water Resources and the Utah Division of Parks and Recreation to file for a change of perfected water right to an instream flow use.[17] The two state agencies may lease, purchase, or receive as donation water rights from other uses and transfer the rights to instream. The statute, however, has had little effect on improving stream flows. Between the two agencies, only a few instream flows rights have been established (Hawkes 2006).

In 2007, the state legislature created a program that provides a private sector alternative to state control of instream flows. The passage of HB 117, a 10-year pilot program, permits private entities such as TU to lease water from farmers and ranchers to provide flows for fish. The legislation is the product of years of work on the part of environmental advocates, policymakers, and water users around the state. The pilot program is similar to the one implemented in Montana in 1995, which proved to be successful and has since been made permanent.

Washington

The protection of instream flows in Washington stems from a mix of state regulatory control and markets. In 1967, the Minimum Water Flows and Levels Act (Ch. 90.22 RCW), authorized the state's Department of Ecology (WDOE), to "establish minimum water flows or levels for streams, lakes or other public waters for the purposes of protecting fish, game, birds or other wildlife resources, or recreational or aesthetic values of said public waters whenever it appears to be in the public interest to establish the same".[18] The 1967 law was largely superseded by the Water Resources Act of 1971, which expanded the authority of the WDOE, to set base flows in all perennial streams in the state. Along streams not fully appropriated to existing users, the WDOE could set aside water instream that would be protected from future uses. In addition, streams could be closed to new appropriations if it was determined that any new diversions would reduce flows below the minimum or base flows. As of 2008, the WDOE had established minimum flows or stream closures in 25 of the state's 62 Water Resource Inventory Areas—regions that serve as the geographic basis for the WDOE's basin management and instream resource protection programs (Washington Department of Ecology 2008).

Minimum flows were effective in protecting streams with sufficient water, but incapable of restoring flows to already dewatered streams, often where fish populations were at greatest risk. To help address this shortcoming, the state enacted the Yakima Basin Trust Water Rights Act in 1989, and the Water Resources Management Act in 1991, which created the first opportunities to acquire water for instream uses. The laws established a trust water rights program through which unused or conserved water could be protected from relinquishment by donating, selling, or leasing water to the state. The water was then protected for instream or other beneficial uses.

The trust water rights program paved the way for subsequent market based initiatives to address the state's overallocated and dewatered streams. In 2000, the legislature created and funded a pilot acquisition program to improve stream flows in four target basins. In 2001, during the state's second-worse recorded drought, the WDOE entered into 21 leases with farmers to maintain flows for fish along key stream reaches (Adelsman 2003). In 2003, in cooperation with the Department of Fish and Wildlife and the Washington Conservation Commission, the WDOE expanded the pilot program and formally launched the Water Acquisition Program. The goal was to restore stream flows for salmon and trout in 16 target watersheds, most impacted by low flows. Water right holders in these watersheds can lease, sell, or donate all or a portion of their water rights to the state. The rights are then held in trust by the state for the purpose of protecting flows instream.

The trust water rights legislation also created opportunities for private entities to improve flows through market trades. The Washington Water Trust, formed in 1998, is a private non-profit organization that employs innovative, market-based solutions to acquire water in critical stream channels where even small amounts can restore flows sufficient for migratory fish species. Similar to Oregon, in order to legally protect acquired water instream, rights must be held in trust by the state. Despite this limitation the Trust has been widely effective in acquiring water to meet instream flow demands.

Additionally, the Trust continues to expand its role in protecting flows through new and innovative programs. It is currently working with the WDOE in the development and implementation of a mitigation program in the Walla Walla Basin—the Walla Walla Water Exchange—designed to offset the impacts on stream flows from new groundwater withdrawals. Outdoor water users between

May 1 and November 30 must mitigate "bucket for bucket," meaning the amount of water withdrawn from a new well must be offset with the addition of the same quantity of water to the stream, aquifer or drainage zone impacted by that well (Washington Water Trust and Washington Department of Ecology 2007).

New well users may mitigate by acquiring a water right from another user then applying the water to instream use; however, the acquisition process can be costly and time consuming (Washington Department of Ecology 2007). As an alternative, for a set fee,[19] mitigation credits can be acquired directly from the Exchange. Prior to the implementation of the program the WWT began acquiring water rights from willing sellers to effectively "seed" the Exchange. Doing so reduces the transactions cost to new developers while ensuring there is sufficient mitigation water to offset new withdrawals.

Wyoming

Protection of instream flows in Wyoming has proven more challenging. Wyoming reformed its law in 1986, recognizing instream flows for the establishment or maintenance of new or existing fisheries as a beneficial water use.[20] In addition the law provides for the appropriation of water for instream use; however, "[n]o person other than the state of Wyoming shall own any instream flow water right."[21] Under this law, the Wyoming Game and Fish Department is responsible for identifying priority streams, performing studies, and making flow recommendations to the Water Development Commission, which in turn applies to the State Engineer's Office for an instream flow right.

The law was challenged, unsuccessfully, in 2000, when the city of Pinedale voted unanimously to transfer up to 4,800 acre-feet of storage water to instream flows in Pine Creek, which runs through town. The water would have been released during certain times of the year to enhance flows for local trout; however, the state engineer denied the request, citing state law that requires instream flow rights to be held by the state, not an individual or municipality (Benson 2006).

Although limited, the law does provide for market trades of water for instream use. The state may purchase, or receive as donation, water rights intended for instream use. Although private entities are prohibited from holding an instream flow right, they may donate all or a portion of their rights to the state to be used instream; however, in the more than 20 years since the passage of the 1986 law, no acquisitions nor donations for instream use have occurred.

Private entities or individuals may petition the state to appropriate water along suggested stream reaches in order to protect flows for fish and wildlife. As of 2007, 64 of more than 100 applications have been approved, protecting flows on 447.4 miles of streams. The approval process, however, can be slow. For example, the applications of recently approved flows were submitted back in 1991, taking 17 years to complete. In the words of Wyoming Senator and economist Cale Case, "the law has been a dud" and he has described it as "overcomplicated, anti-property and narrowly administered" (Case 2003).

Markets for Instream Flows

The benefits of removing legal obstacles to instream flow markets are clear. Changes to state water laws and a gradual expansion of the prior appropriation system to include provisions for instream flow protection have paved the way for market trades of environmental and recreational flows. Between 1987 and 2007, state, federal, and private entities acquired more than 10 million acre-feet of water for instream purposes through short- and long-term leases, donations, and permanent transfers. During this period, total expenditures for instream flows in the 11 conterminous western states neared $540 million,[22] and market activity continues to expand (see Figure 7.1).

Market trading of water rights has been driven largely by state and federal acquisitions to restore flows for threatened or endangered species, to improve or maintain water quality, and provide flows for fish and wildlife habitat. Between 1987 and 2007, federal acquisitions accounted for roughly 46 percent of total expenditures and 56 percent of total water acquired. Similarly, state agencies' expenditures made up roughly 46 percent of total and 34 percent of total water transacted.

In most states, the legal institutional setting continues to limit or even prohibit private entities from acquiring water directly. And states remain reluctant to move away from centralized control of instream flows and permit private demands to determine how water is allocated among offstream and instream uses. As a result, demands for flow restoration are still largely met, not through economic channels, but through political means, a process that "is often slow, contentious, and expensive" (Sterne 1997).

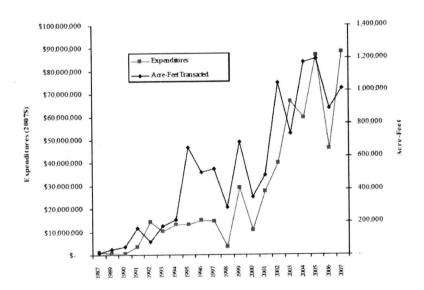

FIGURE 7.1 Instream flow acquisitions, 1987–2007

To the extent that water cannot be allocated by its residual claimants, this centralized approach depends on wise administrators who must know the appropriate amount of water to withhold from private appropriation and how much should be reallocated to maintain instream flows. This approach is fraught with problems of governmental failure (see Chapter 2). When prices do not exist, values are difficult to determine; when voluntary trades do not determine use, opportunity costs are ignored; and when the political process is used to determine allocation, well-organized special interests tend to preempt diffuse general interests. Even where administrators or legislators are not subject to political pressures, it is unlikely that they will have enough information to make efficient tradeoffs among water uses. In the absence of prices, valuing the water in alternative uses is difficult, if not impossible.

Efficiency is further impeded because administrators and legislators do not have to face the opportunity costs of their actions. When water is reserved for instream uses so that consumptive uses are precluded, the decisionmakers do not bear the cost of the foregone alternative uses. To complicate the process, agricultural, mining, industrial, and municipal users will compete with instream users to make their political voices heard. There is no guarantee, however, that political expressions of preference will mirror economic values. As long as existing consumptive rights must be purchased, state revenues will have to be provided. Nevertheless, it is important to remember that the state budget is a common pool, and individuals seeking revenues to purchase water rights will still be able to ignore the full opportunity costs of their actions (Baden and Fort 1980).

Political decisionmakers determine future as well as present water use, and in order to do so, they must be able to accurately predict future values of water in a changing world. Once water is reserved for instream uses, it is difficult to change the uses in the political process. The fear that instream flows will prevent future development of water resources is justified where flows are designated or reserved by the state. Where instream flows are embodied in freely transferable water rights, flexibility both to and from diversion uses is more likely.

Despite continued attachment to governmental control of instream flows, the institutional setting is changing in a way that allows for greater private allocation through markets. Private entities such as water trusts and conservancies now play a critical role in many aspects of stream flow protection and water marketing from education, restoration work, and research to direct water acquisitions that improve instream flows. In addition, they work with state policymakers to remove obstacles to markets and assist in completing trades between governmental agencies and willing sellers. Although private expenditures comprised roughly 8 percent of the total from 1987 to 2007, they have completed more than 1,100 leases, 100 permanent transfers, and received nearly 500 donations—in total, more than double the number of transactions completed by all federal and state agencies combined (see Figure 7.2). And the number of trades completed by private entities continues to increase compared to the number completed by federal and state agencies.

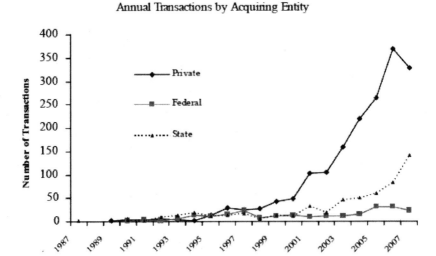

FIGURE 7.2 Historical trends in instream flow transactions by federal, state, and private entities

The evidence suggests that where legal obstacles to establishing instream flow rights are removed, private provision of instream uses will develop and even flourish. In states permitting private acquisitions, more than 2,300 transactions have occurred since 1987 compared to fewer than 150 in Arizona, Colorado, Idaho, Utah, and Wyoming combined. Even in states that may prohibit direct private acquisitions, groups such as TU and the Colorado Water Trust have had notable success in restoring stream flows by coordinating with state agencies and water rights holders to facilitate voluntary transactions. In Idaho, TU worked with irrigators along Rainey Creek, a tributary to the South Fork of the Snake River, to obtain the first water donation to the Idaho Water Supply Bank for instream flow protection. In 2007, The Colorado Water Trust acquired consumptive water rights along an important section of Hat Creek, that provides resting and refuge habitat for local brook and brown trout populations. The water rights were donated to the Colorado Water Conservation Board to be converted and held by the state for instream flows.

Such efforts to leave water in streams and improve stream systems, if they are to be successful, require some security of rights to instream flows. The fact that individuals and groups are taking the initiative to maintain and enhance streams for their instream flow amenities, even when rights to the flows are less than certain, is good indication of the ability of markets to facilitate the demand for instream flows. However, these efforts remain limited in the face of the institutional obstacles to private rights in instream flows.

The motivation for acquiring water for instream purposes is to provide flows for the production of environmental or recreational amenities demanded by consumers. When the gains from ownership of instream flow rights accrue directly to the user, the amenities are more likely to be produced and protected over time, providing a clear environmental and economic gain. Conversely, when water rights are acquired and held by the state for the public, the state is neither the direct beneficiary nor user, and thus has little incentive to ensure the environmental or recreational "products" are produced and maintained over time. To ensure the state will provide such amenities is dependent upon the ability of the state to police itself on behalf of the public, an "unlikely prospect" (Pilz 2006).

Conclusion

It is commonly believed that markets in instream flows cannot work because of the public good characteristics of free-flowing streams or because instream flow interests cannot compete in a market with those who divert and consume water. State institutional structures dealing with instream flows reflect this misconception. By refusing to allow private entities to hold instream flow rights, and by holding on to obsolete elements of the prior appropriation doctrine which limit private allocation decisions to only offstream water uses, states have thrown obstacles in the way of instream flow markets. The result is governmental control of instream flows that is fraught with rent seeking, constraints on future development, and limited budgets. States have established only a minimal number of instream flows relative to the total capacity of watercourses in the West. Though western states have taken great steps toward markets, institutional obstacles continue to inhibit market solutions to instream use conflicts. With these barriers removed, we could move a long way toward efficient allocation and use.

Ironically, some of the opposition to instream flow markets comes from the most likely beneficiaries: agricultural and environmental communities. Farmers are afraid that markets will allow environmentalists to buy up enough water to drastically change traditional agriculture and dry up local communities; and environmentalists often do not believe fish and wildlife can compete in a water market with consumptive users. Neither position is realistic.

The agricultural industry should recognize that some reallocation of water from consumptive uses to instream flows is inevitable. The question is how the reallocation will take place. It may be achieved voluntarily through market transactions where current users will be compensated for conserving and transferring water to environmental uses. Alternatively, it can be reallocated by involuntary, uncompensated state regulation, the public trust doctrine, or the Endangered Species Act. If reallocation comes through market transactions, the amount of water actually reallocated is likely to be smaller than if reallocation comes through uncompensated takings. In the latter case, the demands are insatiable. In the former, they are constrained by the pocketbooks of the demanders.

Environmentalists need to understand the inconsistency of their argument that fish and wildlife cannot compete in a market with consumptive water users such as agriculture. On the one hand, they argue that the agricultural industry, which uses most of the water in the West, contributes only a small percentage to western states' economies. On the other hand, environmentalists claim they cannot compete in a market with agricultural water users. It is true that agriculture is a struggling industry and agricultural water prices are often low because of subsidies. If water is not valuable in agriculture, then environmentalists and recreationists should be able to purchase a lot of water for little money. Indeed, most data show that a small percentage of water transferred from agriculture would yield significant flows for instream and other uses. Couple this with the fact that most environmentalists and recreationists have above average wealth and it seems unlikely that they would not be able to compete for water in the marketplace.

In short, markets have failed to cause the economic and cultural upheaval feared by their opponents. Instead, these markets have demonstrated that resources can be devoted to environmental goods without having to rely on state restrictions or regulations that attempt to dictate water use among competing users. Moreover, instream flow markets are a testament to the capacity of the private sector to provide what many argue are public goods. With increasing demand for environmental and recreational amenities, policymakers should be less concerned with how those demands can be met through governmental actions and instead devote their efforts to removing the very obstacles that stand in the way of those who demand instream flows.

8

GOOD TO THE LAST DROP

Water quality is a worldwide problem caused by competing demands on the precious resource. Some people want to use streams and lakes for domestic consumption, some want to use it for recreation, and some want to use it for disposal of effluent. If each of these uses is sufficiently small, each demand can be met without conflict. However, when those wanting to use water for disposal reduce quality enough that the water is not fit for consumption, conflict arises. The policy question, therefore, is whether the conflicting demands can be resolved using markets or whether regulation is necessary.

By and large, policies to improve water quality have focused on regulation, though regulations have not always been easy to put in place or to enforce. Especially in the developing world, releases of effluent from industrial facilities, otherwise known as point source discharges,[1] are largely uncontrolled. In Chile's Maipo River basin, for example, only a small percentage of waste discharges are treated. The same was true of the Rhine River in Europe until the 1980s, when discharges were reduced and treatment improved (World Resources Institute 1990, 163–64). In 1969, the Cuyahoga River in Cleveland, Ohio, was so full of chemicals that it caught fire (Adler 2002). In the late eighteenth and early nineteenth centuries, the Nashua River in Massachusetts periodically changed color from the dye discharged by paper mills located along its banks.

Nonpoint discharges from sources such as agricultural runoff from crops and livestock have proved even harder to control. Though these have leached into surface water and groundwater for thousands of years, improved agricultural productivity has increased pesticide and fertilizer use and other effluent significantly since World War II. As a result, runoff of nutrients from agricultural production causes one of the most widespread of all water quality problems (Selman and Greenhalgh 2009), particularly in developing nations where fertilizer consumption is expected to increase 17–40 percent between 2002 and 2030 (Food and Agriculture Organization 2000).

But agriculture is not the only source of nonpoint discharge. In many developing countries, urban sewage often is untreated. As a result, diseases associated with filthy water kill an estimated 1.8 million children each year (United Nations Development Programme 2009). Cities contribute to water pollution when stormwater drains off streets and overflows sewage systems, when oil spilled by residents on in their driveways washes into storm sewers, and when urban lawns are fertilized (Paul and Meyer 2001).

In the United States, water quality has been governed by the 1972 Clean Water Act (CWA).[2] The CWA unrealistically called for total elimination of the discharge of pollution into the nation's streams and lakes by 1985. Because that goal was impossible to reach, it has been criticized by economists for ignoring costs. Nonetheless, calls for cost–benefit analysis as a way of identifying the "optimal" level of pollution have been unheeded and alternative mechanisms such as markets have received little attention.

Like other water problems, water quality degradation results from unclear rights to water use. As with water quantity problems, well-defined and enforced property rights can improve water quality. The most obvious way is through liability rules, which, as we shall see below, have been used by the common law to hold those who degrade water quality accountable for the costs they impose on others. Another way is to create rights in the form of tradable permits that limit the total allowable discharge of effluent, but allow emitters to meet the discharge limits through trading. Instead of these more market-oriented approaches, however, regulation has been the norm.

Implementing market solutions to water quality problems requires determining who has what rights to use water and how to enforce those rights once they are identified. In short, this means that rights must be defined and enforced. Before passage of the CWA in 1972, common law courts were defining property rights to land and water in lawsuits based on trespass, nuisance, and other torts.

Though the no-discharge goal of the 1972 CWA embodied the notion that no one has the right to pollute the nation's waters, the 1977 amendments stepped back from this unrealistic goal, by shifting the emphasis to establishing water quality standards and attempting to attain them by regulating point sources of emissions (Brown and Johnson 1984, 952). Subsequently, the 1987 amendments to the CWA shifted the focus from point sources to nonpoint sources, thus opening the door for water quality trading and pollution credits. Even though environmentalists protested the use of tradable emission credits on the grounds that they create rights to pollute, the cap-and-trade approach to improving water quality has gained traction.

Achieving meaningful reform in water pollution regulation will require innovative institutional changes. This chapter explores the merits of common law doctrines to help define property rights to clean water. Where the costs of using the common law are too high to establish rights to clean water and liability for polluting it, tradable water quality permits, especially for nonpoint sources, have the potential to go beyond the current regulatory process.

Property Rights through Common Law

Prior to 1972, most water quality issues in the United States were governed by common law. The common law evolves from the bottom up as plaintiffs and defendants take their conflicting claims over water use to courts for judges and juries to decide who has what rights (Black 1983, 144). With regard to water quality, individuals alleging damage from emissions seek compensation or injunction, or both, to stop effluent discharge. Historically, nuisance, negligence, and trespass law were used to establish rights and resolve conflicting claims (Davis 1993).

Common law focuses on resolving conflicts based on protection of property rights.[3] Just as owners are protected from unreasonable interferences with the use and enjoyment of their property (nuisances), or from unauthorized physical invasion of their property (trespass), riparian and appropriative water rights holders are protected from a diminution of water quality. If any of these rights are violated by another's effluent reducing water quality, the common law provides a remedy. However, it takes more than petty inconveniences to support a plaintiff's claim. Liability can be imposed only with sufficient and credible evidence of harm. Riparian water users, for example, are required to prove that the use of water as an effluent disposal medium is unreasonable in relation to the stream's other users. Nuisance law requires plaintiffs to prove not only that the defendant's conduct is unreasonable, but that the harm it caused plaintiffs is substantial. Trespass requires proof that an intrusion was unauthorized and either intentional or negligent or that it involved abnormally dangerous activity (Buchele 1986, 612–13, 616–17).[4]

In contrast with statutory law, the common law relies on evidence of injury in a specific situation to determine the appropriate level of water quality. If an effluent discharge causes no injury, it may be acceptable; but if it causes injury, the discharge must be ended or reduced to a non-harmful level and damages must be paid (Meiners and Yandle 1994b, 7–8). By addressing pollution in specific cases, the common law holds dischargers directly accountable to those they harm.

Under the common law system, individuals whose property rights are violated by effluent into the water have standing in court to claim redress for past damages and to stop future damages. For example, in Great Britain, ownership of fishing rights enables anglers to protect water quality via common law nuisance and trespass actions. Because a healthy salmon or trout fishery requires clean water, owners of fishing rights have an incentive to monitor water quality. Fish Legal (previously known as the Anglers' Conservation Association, founded in 1948) was formed to fight pollution on behalf of angling clubs in England and Wales. In its more than 60-year history, it has lost only three of thousands of cases and recovered millions of pounds from polluters (Fish Legal 2010). Significantly, the polluters they defeat often have injunctions placed on them which stop further polluting activities. Because the English common law provides protection of fishing rights and forces polluters to pay and/or cease polluting, few pollute the same stretch of river twice (Bate 1994, 14).

In the United States, common law cases have also been successful in addressing water quality disputes. In a 1906 case, a riparian landowner successfully sued the

International Paper Company for polluting a stream that ran through his land, ruining his fishing business and reducing the value of his property.[5] In *Sammons v. City of Gloversville,*[6] a New York court issued an injunction against a town's sewage disposal practices. The court found that the town's practice of emptying its sewers into a creek that flowed through a farmer's land caused filth to accumulate on the creek's bed and along its banks, and held that this act constituted a trespass of the farmer's property rights. The town was enjoined from further sewage disposal into the creek in spite of the public necessity of the sewage works and the great inconvenience and cost that could result from the injunction. The court specified that the injunction would not become operative for a year, however, giving the town time to establish a different sewage system or obtain legislative relief (Brubaker 1995, 32).

Another 1906 case, *Missouri v. Illinois,*[7] exemplifies the proof requirement in common law cases. In that case, Missouri tried to enjoin Illinois from allowing Chicago to dump raw sewage into the Desplaines River that ultimately flowed into the Mississippi River, the source of St. Louis' drinking water supply. Missouri alleged that the sewage carried typhoid-causing bacteria downriver to St. Louis, resulting in almost 200 deaths per year. However, the evidence in the case showed that the bacteria could not survive the trip. Because the evidence did not support Missouri's claim, the injunction was denied (Meiners and Yandle 1994b, 5).[8]

More recently, in 1983, the state of New Jersey held two corporations liable for creating a public nuisance by dumping mercury-laden waste material on its property and into a creek. The corporations were ordered to abate the nuisance by cleaning up the mercury which had been dumped in the waterway, by ceasing to emit mercury, and to pay damages to the state.[9]

The requirement that harm be proven is both a strong point and a weak point in the common law approach to pollution control. On the one hand, proof requirements protect accused polluters from frivolous lawsuits and claims that can be very costly. On the other hand, it can be an obstacle to recovery in legitimate cases. For example, when multiple sources contribute to the pollution of a stream, it is difficult to pinpoint responsibility and to prove a causal connection between the defendant's actions and the damage suffered by the plaintiff. It can also be difficult and costly to prove harm from pollution when a large number of individuals each suffer only a small amount of damage. Such victims tend to become free riders on legal actions brought by someone else against the polluter(s). Class action suits, of course, can greatly reduce the magnitude of this problem, and associations such as Fish Legal in the United Kingdom, may be formed to help overcome it. Property owners' associations are common among landowners, and could be common among water owners if property rights to clean water were well defined and enforced.

The transaction costs associated with using the common law to control pollution decline as our knowledge of the effects of pollution advances, making it easier to prove the cause and extent of injury. With knowledge comes responsibility in the form of tougher liability standards on polluters. According to Meiners and Yandle (1994b, 4),

[T]he common law, not restricted by statutes, might have provided more ecologically and economically sound pollution control than has occurred under the CWA [footnote omitted]. Advances in pollution control technology and in understanding the consequences of pollution as well as changes in society's attitude about the acceptability of pollution would have encouraged further development of a common law standard of strict liability for polluters—which is close to what existed before the CWA.

In addition, the common law provides for injunctions against pollution activities so as to eliminate harm, even if it means a polluter must be shut down. Under the CWA, on the other hand, dischargers who violate their permits are merely ordered to comply with the law and possibly pay a fine. Penalty amounts are usually trivial when compared with the cost of compliance and the damage that may have been inflicted on others. Moreover, the penalty is paid to the government, not to the damaged individuals (Meiners and Yandle 1994b, 8).

A criticism of the common law approach as opposed to a legislative–regulatory approach is that the former is sporadic and ad hoc. When individual judges in different areas of the country decide cases based on unique facts and state law, the legal standards that develop may be difficult to apply in other cases. Moreover, water pollution often involves highly technical issues that can be beyond the competence of courts (Buchele 1986, 5–6).[10]

Perhaps the most often raised critique of the ability of the common law to address pollution problems adequately is the seeming lack of pollution control prior to passage of the CWA in 1972. If the common law works so well, why did the Cuyahoga River catch fire in 1969? Why did the victims of pollution not sue polluters and win large damage awards, thus deterring other polluters? The answer is not entirely clear, but we can speculate about some possibilities. First, it is costly to coordinate individuals who may be harmed to bring suit against the polluter. Second, especially in the case of nonpoint sources of pollution, it can be costly identify who is responsible for the harm. Third, even before the CWA, various federal and state statutes reduced the incentives of private parties to enforce common law restrictions on water pollution (Beck and Goplerud 1988, 204.2). Fourth, the common law was not perceived to be adequate to the task of curtailing environmental harms in the 1960s and early 1970s.[11] And finally, a growing body of evidence suggests that the demand for environmental quality is positively related to income (e.g. Dasgupta et al. 2002). Without a doubt, it was during the late 1960s and early 1970s that Americans began to express their demand for cleaning up the environment on a broad scale.

From Property Rights to Regulation

The Clean Water Act was passed in 1972 with the overall objective of restoring and maintaining the chemical, physical, and biological integrity of the nation's waters.[12] On the way toward achieving that objective, the Act established the goals of rendering the nation's waters fishable and swimmable by 1983 and eliminating the discharge of water pollutants by 1985.[13] The process for achieving those goals focused on controlling water pollution from point sources by regulating what treatment was required before effluent could be discharged. The act directed the Environmental Protection Agency (EPA) to establish nationally uniform effluent limits based on the best available technology (BAT) for reducing effluent and to issue discharge permits specific to what could be achieved using those technologies (Toft 1994, C6). The CWA also required the states or the EPA to establish water quality standards, but states were slow in doing so both prior to and after adoption of the Act (Bonine and McGarity 1984, 426).

At first blush, technology-based effluent regulations appear easier to implement and enforce than water quality standards. The latter would require working backward from a polluted body of water to determine which point sources are responsible for quality deterioration and which should be treated to attain a specified water quality. Once an agency sets the BAT standard, it need only observe whether that technology is in use.

Amendments to the CWA in 1977 shifted emphasis from BAT requirements to water quality standards. They required the EPA and states to establish quality standards for water bodies regardless of whether the technology existed to achieve compliance. Establishing such standards potentially offered an improvement over BAT because water quality standards are less subjective and therefore more transparent. Moreover, performance-based quality standards set the level of effluent and allow dischargers to find the most effective way of achieving that level. The outcome is what counts (see Meiners and Yandle 1994b, 10).

Though point source discharges have been significantly reduced through regulations, the regulatory process has missed a significant and growing part of the nation's water quality problem, namely nonpoint discharge sources. The 1987 amendments to the Act began the first coordinated assault on nonpoint sources, currently the chief cause of impaired surface water quality (U.S. EPA 2008b). Nonpoint sources of pollution include erosion from cultivated land, runoff containing pesticides, herbicides, and fertilizers from farms and suburban lawns, and toxic metals and chemicals from streets and highways (Riggs 1993, ii; World Resources Institute 1993, 40). Though some have advocated treating nonpoint sources the same as point sources (Davidson 1989; Foran et al. 1991), the difficulty of detecting and monitoring pollution from these diffuse, ill-defined sources, as well as the seemingly overwhelming number of human activities that would be implicated by controls on nonpoint source pollution, has so far prevented application of point source regulations to nonpoint sources.

Section 319 of the 1987 amendments therefore established the Nonpoint Source Pollution Management Program, requiring states to develop programs for assessing and managing nonpoint source pollution (Malik et al. 1993, 959). Under the program, states, territories, and tribes receive funding to help implement programs and projects designed to reduce nonpoint source pollution. Since 1990, states have received more than $3 billion in federal grants (U.S. EPA 2010a), primarily for assistance in developing and implementing best management practices. An example is a program called "Operation Green Strip," sponsored by the Monsanto Company, to encourage farmers to plant grass buffer areas along streams and wetlands to filter runoff and prevent soil erosion (*U.S. Water News* 1994).

Another source of nonpoint pollution which has been targeted in subsequent proposals for CWA reauthorization is combined sewage overflow caused by the combination of sewage and storm water runoff in municipal sewage systems. Combined systems can overflow during rainstorms, sending untreated sewage into streams and rivers (Ember 1992, 19). Most reauthorization proposals provide grant money to states to help them overhaul combined sewage overflow systems.[14] For example, the Chesapeake Clean Water and Ecosystem Restoration Act of 2009 seeks to amend the CWA to provide an additional $1.5 billion to local governments in the Chesapeake Bay watershed for projects to reduce stormwater runoff. An additional $625 million would be used among the six states and the District of Columbia (all within the watershed) to fund implementation, monitoring, and assistance grants (Chesapeake Bay Foundation 2009).

Despite no major amendments to the CWA since 1987, there has since been considerable legislative activity. The House of Representatives introduced a comprehensive reauthorization bill (H.R. 961) in 1995 that would have, among other things, expanded funding, created new programs, revamped the federal wetlands protection program, and amended coastal nonpoint source pollution programs. The proposals were widely opposed by industry and agriculture as unreasonably onerous and environmental groups as insufficient. Although the bill was not taken up by the Senate, it set in motion a number of changes to the CWA in subsequent sessions. In particular, measures have been enacted to increase funding for water infrastructure projects, expand existing programs, create stronger controls on nonpoint source pollution, and provide resources to develop and expand watershed management programs.

Managing water quality on a watershed basis, as opposed to political boundaries, is another approach that has generated significant interest, particularly as nonpoint source pollution has come to the forefront. Watershed management involves addressing the cumulative effects of all activities that generate pollution within a water basin, including point and nonpoint sources.

An example of a watershed management approach is the cleanup of Chesapeake Bay. In 1983, a cooperative public–private effort to restore and protect the bay was launched. The Chesapeake Bay Program involves three states, the District of Columbia, and the EPA in coordinating efforts to reduce nutrient and toxic loadings into the bay and restore native vegetation that is essential to the estuary's health. The state of the bay has significantly improved since the program's inception, but nonpoint source pollution remains the number on cause of pollution in the bay.

Not surprisingly, watershed management has not been embraced by everyone, despite its obvious benefits. States are wary of it because they fear the primacy in water quality management they have enjoyed under the CWA may be undercut by federal watershed mandates.[15] Farmers fear it as a first step toward a permit system for agricultural nonpoint sources. On the other hand, environmental groups embrace the idea of watershed management. They would like to see it become mandatory rather than voluntary, and because they fear the approach would allow states to ease up on point source dischargers, environmentalists insist that the EPA should have strong oversight of watershed programs (Rubin et al. 1993).

As we will see below, watershed management is a necessary step in the direction of pollution credit trading. In a basin where both point sources and nonpoint sources contribute to pollution loading, trading within and between both groups provides multiple opportunities for economically efficient pollution reduction. The House of Representatives acknowledged as much when it introduced H.R. 961, the Clean Water Amendments of 1995. This bill included a provision encouraging states to use voluntary watershed management programs. Under it, states would be granted flexibility in issuing point source permits under watershed management plans, which in turn would facilitate pollutant trading.

Proposals to impose fees on point source discharge permits have also been included in CWA reauthorization bills. Supporters of fees see them as a means of funding the costs of permit administration, as well as other CWA programs. While industry is not opposed to the concept of fees to underwrite administration of the CWA's permitting program, it is not eager to pay them (Ember 1992, 20).

Wetland regulation is a particularly controversial issue in the CWA reauthorization. Both urban and agricultural development can disrupt wetlands by draining or filling them.[16] While not intuitively related to water pollution control, wetlands are environmentally important because they purify water, protect against floods and erosion, recharge aquifers, and provide fish and wildlife habitat (Hanson 1994, 39). Wetlands adjacent to navigable rivers are included in the CWA definition of "waters of the United States" over which the federal government has jurisdiction.[17]

Under its regulatory authority, the Corps of Engineers and the EPA may limit or prohibit development of property that includes wetlands. With an estimated 75 percent of remaining wetlands located on private property (U.S. Congressional Research Service 2008a), landowners are particularly exposed to land and water use restrictions and they claim that such regulation decreases the value of their property, entitling them to compensation under the 5th Amendment. The 1994 CWA reauthorization bill would have provided some financial incentives for wetlands conservation, but environmental groups and key White House officials opposed compensation measures, claiming they would raise serious fiscal problems. Environmental organizations also argue that protecting wetlands cannot constitute a taking because the importance of conserving wetlands supersedes any individual's economic interests (Hanson 1994, 39). These arguments notwithstanding, the 1995 Clean Water Act Amendments passed the House again and included compensation provisions.

Improvements to water quality under the CWA, its amendments, and other legislation have come at substantial costs. By 1987, U.S. businesses, government, and individuals had spent over $300 billion for water pollution control (Conservation Foundation 1987, 87). According to a cost–benefit comparison published by Resources for the Future in 1990, the average annual benefits of federal water pollution control between 1979 and 1988 were $15 billion, while costs averaged $23.2 billion (Meiners and Yandle 1994a, 77–78). In 2004, the EPA estimated the additional public funding needed to meet CWA standards for the next 20 years would be $202.5 billion, or roughly $10 billion each year (U.S. EPA 2008). This figure is in addition to the estimated $390 billion in private capital costs needed to repair and upgrade infrastructure in order to meet future water quality standards (U.S. Congressional Research Service 2008).

And there is little to suggest that the cost of water quality regulations is likely to go down. Funding provided by Section 319 of the CWA was intended to provide states assistance to implement nonpoint source abatement plans; however, the efficacy of such plans is still unclear (U.S. Congressional Research Service 2008). Nevertheless, President Obama's 2011 budget provides $2.1 billion in grants, sometimes in the form of loans to states, to initiate roughly 800 new clean water projects (Office of Management and Budget 2010).

The high cost of regulation and meeting water quality standards under the CWA underscores the importance of ensuring the resulting benefits match those costs, especially when public monies are being appropriated. However, quantifying the benefits from clean water projects and best management practices has proven challenging, creating uncertainty as to whether funding is going to its highest valued use and maximizing water quality outcomes. Moreover, the inherent political nature of regulatory policies provides an opportunity for misappropriation of funds.

In testifying before Congress in 1999, Wyoming's governor voiced concerns over the efficacy of the CWA and, in particular, the use of resources to address water quality. During the hearing he asked "what is it we are trying to achieve? . . . [And] are we effectively allocating the resources today?" The governor went on to explain that "the authority for states to receive federal money for watershed work required that we declare that a waterbody was functionally impaired— regardless of its actual condition. That misunderstood incentive caused many streams to be mislabeled as impaired. As a result, Wyoming was able to draw down 319 money."[18] Needless to say, such use of federal money suggests that cost–benefit analysis is not a driving force behind these regulations.

Common Law vs. Statutory Regulation

Though the common law did not perfectly control water pollution, it was largely supplanted by passage of comprehensive federal water pollution legislation—the CWA in 1972—which effectively preempted many incentive-based enforcement mechanisms generated by clarifying rights. Although the CWA contains a clause preserving state law and other remedies against polluters, it has been narrowly construed by the U.S. Supreme Court in the face of the comprehensive nature of the act's regulatory scheme.

In *Milwaukee v. Illinois* in 1981, the high court held that federal common law was preempted by passage of the CWA because in passing the Act, Congress filled the field of water pollution control with legislation, removing the need for federal court-made common law.[19] As a result, Milwaukee's sewage disposal into Lake Michigan could not be enjoined as an interstate nuisance so long as Milwaukee complied with its CWA discharge permit. Hence, Illinois was left without a federal common law remedy.

Six years later, in *International Paper Co. v. Ouellette*, the U.S. Supreme Court again limited the power of state common law.[20] Therein it held that Vermont could not apply its own nuisance law to impose liability on a New York discharger located on Lake Champlain, or to require the discharger to comply with more stringent controls than those imposed under a CWA discharge permit issued by New York.[21]

The CWA was enacted to protect people from the effects of pollution, but it actually limits the rights of citizens to seek protection and redress from polluters (Meiners and Yandle 1994b, 7). Under the CWA, regulators establish the acceptable level of pollution when they promulgate effluent limitations and issue discharge permits. This "statutory approach presumes regulators are correct in their knowledge about how much pollution is 'right'" (Meiners and Yandle 1994b, 91). No matter how well intentioned or informed agency officials may be, however, it is impossible for them to determine the optimal level of pollution in every case. They are subject to political pressures and, in promulgating regulations, they set uniform standards that are applied over a broad range of pollution contexts. Moreover, the CWA's technology-based regulations sometimes require pollution control equipment to be used even if the pollutant is harmless. Worse, technology-based regulations can allow emission of a harmful pollutant if the best control technology currently available is inadequate (Meiners and Yandle 1994b, 8). When compared with the common law, which determines the level of emissions on the basis of evidence of injury in individual cases, the statutory method of setting uniform pollution standards breaks the connection of accountability between polluters and citizens affected by pollution.

Not unlike other environmental statutes, the CWA allows any citizen to bring a civil action against a polluter who is in violation of effluent limitations or standards or of the polluter's discharge permit.[22] Citizens may also sue the EPA for failing to enforce the CWA. Citizen suits are limited to enforcing effluent limitations, standards, or permit requirements, and may include imposition of civil or criminal penalties, an order requiring the polluter to comply with its permit and the act, and a possible award of costs and attorney's fees to the prevailing citizen. The CWA contains no provision for any payment of damages to parties injured by the pollution.[23] This limitation reduces the incentives of injured parties to seek redress, which would carry with it the benefit of a general reduction in pollution and environmental damage (Meiners and Yandle 1994a, 91).

The CWA requires individuals who intend to bring a citizen suit to give 60 days' notice to the discharger. If the polluter is able to bring itself into compliance within the 60 days, the suit cannot proceed unless the violation is alleged to occur intermittently.[24] No such grace period exists under the common law. If harm giving rise to a cause of action is inflicted, a suit for damages may be brought within the time period prescribed by the applicable statute of limitations.

In short, the CWA does not provide the same remedies to those adversely affected by water pollution as does the common law. At best, the enforcement provisions of the CWA bring point sources into compliance with their discharge permits, but compliance will not necessarily alleviate the pollution problem or make the harmed party whole. In contrast, the common law clarifies property rights and makes polluters directly liable for damage caused by pollution, giving them the incentive to develop better control technology and methods, and giving injured parties the incentive to sue for damages and an injunction.

The strengths of the common law become evident when set against the regulatory structure of the CWA, as an example from an Oregon lawsuit in the 1990s illustrates. The Stevenson family owned the Knee Deep Cattle Company near Coburg, Oregon, where they were running 1,000 head of cattle that drank from the Little Muddy Creek. When developers Bindana Investments Co. built a recreational vehicle park next to a hotel near the Stevenson ranch, Mike Stevenson wondered if the motel's sewage treatment plant would be able to handle the additional volume. His fears were confirmed when untreated (and unreported) sewage was dumped from the plant into the stream (*U.S. Water News* 1995, 7). Stevenson's cattle began to get sick, and the creek was contaminated more than once. The rancher filed both a citizen suit under the Clean Water Act against the hotel and RV park owners, and a civil suit alleging trespass, nuisance, negligence, and strict liability. The Stevensons estimated their economic damages to be $350,000 and the decrease in their property values to be $568,000. They were also seeking $2 million in punitive damages.

The CWA citizen suit was dismissed by the federal district court on the ground that Oregon's Department of Environmental Quality (DEQ) had diligently prosecuted Bindana for violating their permit.[25] When the sewage dumping was reported, the DEQ entered into an agreement with the defendants requiring them to upgrade their facility. Although the defendants violated interim discharge limits established in the agreement and were assessed penalties, the court held the agreement was a shield against further enforcement liability under the CWA citizen suit provision. The court also stated that citizen suit plaintiffs are not entitled to a personalized remedy.[26] The Stevensons appealed to the 9th Circuit Court of Appeals because of the defendants' continuing violations of the terms of their discharge permit and the agreement with DEQ. The 9th Circuit Court reversed the district court decision, on the basis that the ongoing actions by the DEQ did not constitute "diligent prosecution" against Bindana; therefore, the citizen suit was not precluded.[27]

The Market Approach to Water Pollution

Given that the common law has largely been eroded and replaced with a regulatory approach that is costly and inefficient, the question is whether there is an alternative incorporating regulation with tradable property rights. One possibility is to establish water quality standards for a river basin and allow the users of that basin to use market mechanisms to meet the standards collectively at the lowest cost. In recent decades water quality trading programs have received considerable interest, not just from economists, but from state and federal policymakers and some environmental groups (King and Kuch 2003).

Though the CWA does not specifically authorize markets in water quality credits, credit markets arise under the act's directive to states to establish plans to control nonpoint source pollution (Bartfeld 1993). States have been drawn to incentive-based market schemes for controlling nonpoint sources as a positive alternative to the CWA's technology-based regulatory approach. As we have seen, regulations can reduce pollution, but at costs that may far exceed benefits. Regulations based on uniformity may make it easier to enforce rules (Yandle 1994, 187), but the cost of pollution control to individual point sources varies with the nature of their locations and operations. By failing to take this variance into account, uniform regulation such as the CWA's technology requirements is far more expensive than it need be.

Moreover, as many point sources have already achieved significant effluent reductions, the cost of achieving additional controls is now much higher than the cost of implementing nonpoint source controls. At the margin, further point source reductions would contribute less toward water quality in many basins mainly because point source effluent discharges now constitute a smaller percentage of pollutant loadings than nonpoint sources (U.S. EPA 2008a). For example, prior to implementation of a trading system in North Carolina's Tar-Pamlico basin in 1989, more than 80 percent of the nutrient pollution entering that watershed came from nonpoint sources (Riggs 1993, I; Hall and Howett 1994, 27).

In short, nonpoint source emissions must be addressed if further water quality improvements are to be realized. By their very nature, however, nonpoint sources are dispersed and numerous and therefore not as amenable to regulation (Willey 1990, 575; King and Kuch 2003). In addition, attempts to regulate agricultural and municipal emissions have met significant political opposition (Battle and Lipeles 1993, 422).

Because of the inherent limitations of controlling nonpoint source emissions through direct regulations and the need for a more cost-effective strategy, the EPA, in 1996, issued a Draft Framework for Watershed-Based Trading (U.S. EPA 1996). The document was developed to provide technical assistance for the design and implementation of trading programs and guidance as to how the EPA intends to regulate trading within the context of the CWA. Despite some shortcomings, the document led to a number of trading programs and broadened the interest in water quality markets (Copeland 2003). In 2003, the EPA issued a Final Water Quality Trading Policy to "encourage states, interstate agencies and tribes to develop and implement water quality trading programs for nutrients, sediments and other pollutants where opportunities exist to achieve water quality improvements at reduced costs" (U.S. EPA 2003, 1). The number of trading programs in the United

States has since grown considerably. However, because trading activity remains low, the EPA reports that the goals of reducing nonpoint pollution are far from being met (U.S. EPA 2003).

The Mechanics of Water Quality Trading

Markets in water quality represent a "bubble" approach to meeting pollution targets. The bubble theory, which originated under the Clean Air Act, treats sources within a designated area as if they were all under an imaginary bubble. The total allowable level of pollution emitted into the bubble is determined politically, but within the bubble, regulated sources allocate discharges among themselves according to relative economic efficiency.

In the context of water pollution, bubbles can be placed over specific water bodies or entire watersheds within which each point and nonpoint source can be allocated a target discharge level, the sum of which being equal to the bubble. If it is too expensive for a point source discharger to meet his target level, he can buy credits from other point or nonpoint sources that have reduced their pollution levels below their respective target or permitted level. Point and nonpoint sources with lower control costs have the incentive to reduce pollution amounts, thereby creating tradable pollution credits. Higher-cost dischargers would buy credits and clean up less. Either way, the net amount of discharge would not exceed the allowed amount which, again, was established in a command-and-control process (Yandle 1994, 188).

Differences in pollution control costs create incentives for gains from trade, thus enhancing efficiency. If all sources in a basin have identical abatement costs, the net gains from trade would be zero and trading would present no advantage (Riggs 1993, 30). But differences in control costs can be substantial among point sources and between point and nonpoint sources. For example, the cost of meeting effluent targets in Ohio's Great Miami River Watershed over a 20-year period from point source reductions was estimated at $422.5 million, while costs for nonpoint reductions were $46.5 million. In the same watershed, additional point source reductions in total phosphorus were estimated to cost $23.37 per pound compared to $1.08 to $8.48 for nonpoint reductions using certain best management practices (BMPs) (Kieser and Associates 2004). The difference was substantial enough to encourage the Miami Conservancy District (MCD), in conjunction with a number of state and federal agencies, local farmers, environmental advocates, and community watershed groups, to develop a watershed-wide trading program and provide supplemental funding for the implementation of BMPs that could improve water quality at a lower cost. In 2006, the MCD launched a 10-year pilot trading program, the Great Miami River Watershed Water Quality Credit Trading Program, which permits point sources to offset excess phosphorus and nitrogen discharges by financing BMPs on agricultural lands (U.S. EPA 2008).

To create a market in pollution credits it is first necessary to establish target levels of pollutant loadings for individual pollutants in a water body or basin.[28]

Those targets are generally determined by the maximum amount of a pollutant that may enter a water body without violating water quality standards for that particular pollutant; often referred to as the total maximum daily load (TMDL). Once a TMDL or similar framework is established, pollutant loadings can then be allocated across all point sources (waste load allocations [WLAs]) and nonpoint sources (load allocations [LAs]). In general, nonpoint pollution sources are not regulated under the CWA and thus not required to meet LAs, unless generating credits for trading. Point sources, however, are regulated under the CWA and required to have a National Pollutant Discharge Elimination System (NPDES) permit before discharging any pollutant into a water body, as defined by the CWA. With a trading program in place, WLAs can be incorporated into the NPDES permits, allowing point sources to exceed their permitted levels as long as sufficient credits or offsets are obtained. In the absence of a trading scheme, point sources must not exceed their permitted levels.

Before credits can be generated, pollutant loading baselines must be established. In most cases the loadings allocated by the TMDL or other regulatory limits serve as the baselines for individual sources. To generate credits, a source would have to reduce pollutant loadings below its respective baseline—the WLA for point sources or the LA for nonpoint sources if set by the TMDL. Because of the diffuse nature of nonpoint sources and difficulties in measuring changes in effluent levels, state trading programs often specify additional measures that must be taken before credits may be generated from nonpoint sources. Often, certain BMPs must be implemented in order to meet baseline targets, and then only additional changes in management practices can generate credits.

Once baselines and load allocations are determined, it is important that pollution credits are easily tradable. The flexibility by which regulated point sources can offset their discharges through markets is determined by provisions in the state trading program and its permitting authority. A number of trading scenarios are possible, although not always feasible or permitted by the associated program. Trading among point sources is generally the simplest and may carry the lowest transaction costs because of the ease and accuracy of measuring point source discharges (U.S. EPA 2007). When nonpoint sources are the leading cause of water quality concerns, point sources may gain from purchasing credits generated from changes in land management in place of costly technological changes. However, the transaction costs of nonpoint source trades can be higher because of the costs of implementation, monitoring, enforcement, and uncertainty about the effectiveness of BMPs to reduce effluent levels (discussed further below). In any trading scenario, individuals must be free to seek willing buyers and sellers and to consummate trades through efficient administrative channels. Brokers and credit exchanges can play an important role in orchestrating trades and reducing transaction costs.

How pollution credits are generated must also not be restricted by technology-based requirements that dictate how discharge levels are to be achieved. By setting standards and granting participants flexibility in achieving them, markets create a discovery incentive, wherein dischargers will seek the most cost-effective methods of abating pollution to minimize costs and maximize profits. In addition, when they can profit by finding cheaper ways to reduce discharges, they have an incentive to discover and apply new technologies (Riggs 1993, 29; Yandle 1994, 188). Leaving the means of pollution abatement to those with time- and place-specific information about control costs and effectiveness gets the incentives right.

Monitoring and enforcement of discharges under a water quality trading program is no less important than it is under a regulatory system. Under the CWA, the EPA and states rely heavily on polluters to provide the bulk of data on discharges. Not surprisingly, dischargers are reluctant to report their own violations, and regulators lack the budgets or manpower to address all violations even when they are aware of them.

For example, a General Accounting Office investigation of 921 major air polluters officially deemed in compliance revealed 200, or 22 percent, to be violating their permits; in one region, 52 percent were out of compliance (Ackerman and Stewart 1988, 182). Even when illegal polluters are identified, they are not effectively sanctioned. For example, the EPA's inspector general in 1984 found that it was common practice for water pollution officials to respond to violations by issuing administrative orders which effectively legitimize excess discharges (Ackerman and Stewart 1988, 182).

In contrast, markets provide strong incentives for private monitoring and enforcement because property rights have value only when they are enforced. For pollution credits to be of value to their buyers, they must be defendable as well as capable of being divested. The EPA recommends any trade agreement to clearly state the respective responsibilities of buyers and sellers and associated liability in the event of a default or failure of an offset to generate real reductions in effluent levels. By default, the "permittee using those credits is responsible for complying with the effluent limitations that would apply if the trade had not occurred" (U.S. EPA 2003). In the Kalamazoo River Water Quality Trading Program, if a nonpoint source fails to comply with the terms of the agreement to produce pollution credits, the nonpoint source can be held accountable and required to refund any money from selling credits (Morgan and Wolverton 2005). Accountability and enforcement ensure real reductions in pollution levels can be achieved and provide credit producers and consumers incentives to enforce trades.

By requiring compliance with water quality standards and allowing markets to enforce pollution rights, agencies can reduce their monitoring burden. Rather than being responsible for ensuring that numerous pollution sources comply with their discharge permits, water quality markets allow regulators to concern themselves with establishing and enforcing water quality standards which become the binding constraint on the basin or water body and, hence, drive the market.[29] The focus shifts from requiring installation of certain pollution control technology, which may or may not result in clean water, to the actual condition of the environment.

Current State of Water Quality Markets

The number of water quality trading programs in the United States has grown considerably since the first few in the 1980s.[30] The EPA (2010) identifies 48 domestic trading programs in 25 states. Of the 48 programs, 24 had completed at least one trade and four were still under development. Of the active programs, 23 allow point source dischargers to acquire credits from nonpoint sources, and six of those programs also allow point-to-point source trades. There are two programs that permit trading between nonpoint sources—the Grassland Area Farmers Tradable Loads Program in California and Lake Dillon Reservoir in Colorado.

To meet water quality standards, states may choose to develop different trading frameworks depending on the nature of the pollutants and water quality goals. Most concerns over water quality arise from excess nitrogen and phosphorus loadings, primarily from runoff of agricultural fertilizers and discharges from point source wastewater utilities. Not surprisingly, the majority (roughly 75 percent) of trading programs in the United States and in other countries have been developed to address these nutrients. Other programs have been designed to reduce discharges of heavy metals (Clear Creek, CO, and Passaic Valley Sewerage Commission Pretreatment Trading, NJ), and selenium (Grassland Area Farmers Tradable Loads Program, CA, and the Lower Colorado River, CO) (Breetz et al. 2004).

There is also growing interest in programs to mitigate sedimentation of streams, high biological oxygen demand (BOD) levels, and rising water temperatures, particularly in smaller streams. In the Tualatin River basin in Oregon, for example, the EPA and Oregon DEQ implemented a trading scheme that permits point sources to trade water oxygen credits with other point sources, and temperature (thermal load) credits with nonpoint sources and other landowners. Clean Water Services, the primary water treatment utility in the basin and co-developer of the trading program, has contracted with farmers and other landowners to plant trees and shrubs along stream banks, providing shade and reducing water temperatures. Between 2004 and 2008, more than 1.6 million trees were planted, providing shade to 38 miles of stream (Clean Water Services 2008). The cost savings of the program have been significant. The estimated costs for Clean Water Services to meet their water temperature targets through capital improvements, such as installing refrigeration systems to cool effluent, would have been $104–255 million over 20 years, compared to the roughly $12.3 million to fund shade tree plantings (Niemi et al. 2007). In addition to obvious water quality benefits, the new vegetation provides habitat for aquatic and terrestrial species, streambank stabilization, and reductions in runoff and sedimentation from adjacent lands.

The nutrient trading program established in the Long Island Sound of Connecticut and New York is a good example of how credit trading can achieve water quality standards more efficiently than traditional regulations. Growing concerns over water quality in the 1980s prompted the formation of the Long Island Sound Study (LISS), a coordinated effort involving the EPA, Connecticut Department of Environmental Protection (CTDEP), the New York State

Department of Environmental Conservation, and private organizations, and landowners. The purpose of the study was to identify the primary causes of and develop a strategy to address declining water quality in the sound. It found that excessive discharges of nitrogen, primarily from sewage treatment plants, was the leading threat to water quality in the watershed. High nitrogen loading in streams was causing excessive algal growth, which when decomposed by bacteria and other organisms, depleted dissolved oxygen levels. In turn, the low levels of dissolved oxygen or hypoxia was disrupting the normal functioning of aquatic ecosystems within the sound and threatening the viability of local fish and shellfish populations. The findings led to the development of a watershed-based nitrogen TMDL and nutrient trading program that calls for a 58.5 percent reduction in nitrogen loadings by 2014, based on a 1990 baseline level. The EPA estimated that implementing a flexible trading program could reduce the cost of meeting nitrogen loading goals by $200–400 million over 10 years, compared to the costs of new capital improvements. In addition, New York City could save an estimated $660 million in costly upgrades to sewage treatment plants (U.S. EPA 2009).

At its core, the program consists of a statewide cap on the amount of nitrogen discharged and a TMDL which allocates WLAs to 79 municipal sewage treatment plants.[31] The cap and individual load targets are adjusted downward each year, gradually reducing total permitted nitrogen discharges into the sound until 2014 when the target 58.5 percent reduction is supposed to be met. To ensure compliance with annual loading targets and individual WLAs, permitted facilities are required to report total discharges to the state each month and the CTDEP monitors and performs annual audits on all permitted treatment plants. Each year, facilities that exceed their target loadings are required to purchase nitrogen credits to offset all discharges in excess of their permitted levels. Facilities that reduce loadings below their respective baselines generate tradable credits.

The CTDEP, in conjunction with the Nitrogen Credit Advisory Board (NCAB), also oversees the Nitrogen Credit Exchange (NCE), a centralized clearinghouse through which credits are bought and sold. Unlike a bilateral or broker assisted trading framework, the NCE is required to purchase all verified credits, which can then be sold to facilities in excess of their WLAs. Credit prices are determined internally by the CTDEP and the NCAB, based on historic capital costs of reducing discharges (Breetz et al. 2004). For instance, in 2002, the credit price of $1.65 per pound of nitrogen was based on the average historical capital expenditures and ongoing operation and maintenance costs for nitrogen reduction projects of 24 sewage treatment plants (CTDEP 2003). Prices may be further adjusted based on the proximity of reductions (generated credits) to areas of particular water quality concern. The value of reducing discharges tends to increase the closer a facility is to hypoxic zones, and credit prices are adjusted to reflect this. As such, higher prices provide added incentives for reducing nitrogen loadings in areas with the worst water quality (Breetz et al. 2004).

The Long Island Sound credit trading program has been the most active in the country. Between 2002, the first year trading occurred, and 2008, nearly 14 million credits totaling nearly $39 million were bought and sold through the NCE (CTDEP 2009). In 2008, all but one facility was involved in the purchasing or selling of credits. Of the 79 permitted facilities, 49 acquired credits to meet loading targets while 29 sold credits created through nitrogen reduction projects, generating roughly $2.7 million in revenues (CTDEP 2009). The program has also been one of the most effective in terms of producing real improvements in water quality. In a comprehensive analysis of 11 current trading programs by the EPA (2008), only the Long Island Sound program had recorded improvements in ambient water quality. By 2006, nitrogen discharges into the sound had been reduced by 15,500 lbs/day, roughly 50 percent of the reduction needed to meet the final target in 2014 (U.S. EPA 2008).

Despite significant reductions in discharges, the program is not without flaws. Because credit prices are determined internally and not by market forces, a disconnect can exist between the annual supply and demand of credits. In fact, in each year trading has occurred, the NCE has carried either a surplus or deficit of credits, meaning annual loadings into the sound were either under or over the target (CTDEP 2009). Moreover, because credits are acquired by the state for the NCE, taxpayers are left with either a bill for the extra credits or water quality levels below the respective year's target (Powers 2003). Annual fluctuations in loadings could be considered acceptable assuming the end goal is met. However, if the program continues operating as usual, it is unlikely to achieve the sufficient reductions to meet the 2014 target (CTDEP 2009). In hopes of still meeting their target, the CTDEP and NCAB are currently considering revisions to the program that would improve the incentive structure for reductions. Included in their proposals are measures to adjust the price of credits, improve monitoring, and impose stricter enforcement on discharge limits (CTDEP 2009).

Instead of relying on the state to determine prices, allowing credit producers and consumers to negotiate trades directly under an enforced annual loading target would improve efficiency, ensure better compliance to water quality goals, and eliminate unnecessary acquisitions.

Obstacles to Water Quality Markets

Despite growth in the number of trading programs and substantial financial and technical support from the EPA, trading activity has so far been limited to a few programs. According to the EPA (2008), only 100 facilities have entered into trades and 80 percent of all trading activity has occurred within a single program, the Long Island Sound Trading Program.[32] The novelty of pollution trading and difficulties in moving programs beyond the implementation stage to actual trading can, in part, explain the low activity. King (2005, 71) described the state of trading programs as "frozen at an awkward pre-trading stage of development." More importantly, however, are the underlying technical, economic, and institutional barriers that complicate trading, especially as they pertain to nonpoint sources.

Many of the key obstacles in the way of expanding trading center around the complexities of nonpoint source pollution. Unlike tracking emissions levels from the end of a point source pipe, nonpoint sources are diffuse, stretching hundreds of acres across a water basin and varying according to the weather, the lay of the land, land use practices, and a long list of other biological, chemical, and physical factors. Moreover, realized reductions in pollutant loadings from changes in land use practices are often delayed and varied, and may accrue only gradually over time compared to technological changes to point sources that are predictable and can yield immediate reductions in loadings. Because of this variation, it can be difficult and costly to accurately measure, monitor, and enforce discharges from nonpoint sources. This in turn, complicates the establishment of baselines, TMDLs, load allocations, and trading ratios, while creating uncertainty about the efficacy of BMPs and other pollution abatement projects to generate reliable credits.

Perhaps more importantly, the inability to accurately measure, monitor, and enforce nonpoint sources can discourage trading by subjecting credit buyers to additional risks. Nutrient credits from nonpoint sources are determined not by actual effluent loading measurements from individual farms, which would be impractical, but by models that predict how changes in land management will produce reductions in pollutant loadings over time. However, because of the complexities in measuring diffuse pollution sources, there are imperfections in these models and predicted reductions are likely to differ from actual loadings (Abdalla et al. 2007). To a regulated point source facility that must meet certain loading targets, acquiring credits from unregulated sources that may or may not produce sufficient reductions in pollution may be too risky and, in the case of a default, would subject the acquiring facility to risks of noncompliance and potential fines (Selman et al. 2009).[33] As a consequence point sources may be more likely to seek credits from other point sources or implement in-house reduction plans, despite the added costs.

Because of the often large disparity in mitigation costs between point and nonpoint sources, there is a strong incentive to reduce some of the uncertainty in order to facilitate trading with nonpoint sources. That said, however, improving the accuracy with which nonpoint sources are measured introduces added costs, creating a tradeoff between more precise knowledge of discharge levels and transaction costs of making trades. Equivalency factors or trading ratios are frequently used to account for or circumvent some of this uncertainty. In the case of the Chatfield Reservoir Trading Program in Colorado, for example, a 2:1 ratio was established for phosphorus reductions, requiring point sources to secure two units of reductions for every one unit over its permitted level (U.S. EPA 2008). This ratio was designed to provide an environmental safety margin to increase the likelihood that real improvements in water quality will be achieved.

Another problem complicating trading is the incentive structure faced by landowners to produce nonpoint source credits. There are a number of state and federal subsidy programs that provide direct financial assistance to landowners for implementing certain land and water management practices. These programs are intended to produce environmental amenities, including better water quality.

However, in doing so they compete with trading programs by reducing the potential availability of credits for trading (King and Kuch 2003). For instance, a state-funded program designed to reduce agricultural runoff on farms would likely improve water quality by reducing effluent loadings into streams, but at the same time reduces the opportunities for landowners to implement additional nutrient management projects that could generate tradable credits. In essence, landowners who implement subsidized projects are reducing their baseline from which credits could be generated (King and Kuch 2003). In the end, point source polluters are left with lower supply of credits. Where subsidized programs produce measurable improvements to water quality, part of the burden and costs of meeting water quality standards are effectively shifted to taxpayers.

Like any market for goods or services, there are certain transaction costs that can reduce gains from trade and complicate market expansion. In water quality markets, these costs include finding buyers and sellers, negotiating credit prices, verifying credits, monitoring projects over time, and enforcement of contracts. While present to some degree in any trade, these transaction costs tend to be higher when involving credits from nonpoint sources (Abdalla et al. 2007). Because individual nonpoint sources generally provide only a small number of credits, buyers must negotiate trades with multiple parties, increasing their costs of contracting, monitoring, and enforcing trades (Woodward 2006).

The use of brokers, exchanges, or a clearinghouse can help to reduce some of these costs by connecting buyers and sellers through a centralized entity. These third parties may also verify, monitor, and enforce credits over time to reduce some of the risk and uncertainty to buyers and sellers. For example, as part of the Chatfield Reservoir Trading Program in Colorado, the Chatfield Watershed Authority acts as a clearinghouse for pollution credits. Nonpoint sources may deposit credits with the Authority, which then sells them, in a 2:1 trading ratio, to point source dischargers which have exceeded their allowances. To ensure water quality standards are met, the Authority administers a watershed-wide monitoring program that annually assesses phosphorus loadings and eutrophication levels in the Chatfield Reservoir (RNC Consulting 2003).

Conclusion

Ultimately, dealing with water quality requires answering the question: Who has the right to use water for what purposes? If that question can be answered, people with those rights can engage in voluntary transactions between willing buyers and sellers. If individuals have the right to consume clean water, those wanting to use it for waste disposal will either have to purchase the right to clean water or pay damages for the harm they cause. Before enactment of the 1972 CWA, common law incrementally answered this question and, in so doing, issued injunctions or required payment for damages or both.

Under the CWA, however, the EPA and state regulators took responsibility for determining the level of water quality and the method by which that quality would be achieved. Plagued by a lack of information and by the fickle winds of politics, governmental agencies cleaned up some of the nation's waters but at an extremely high cost. The limits of the CWA's regulatory approach bring into relief the strengths of the common law in providing a remedy for damages caused by pollution and in deterring further pollution. While the common law has its costs and inadequacies, it does, at least, determine who has what rights and thereby provides incentives to deal with water quality. The deterrent effect of a few well-established cases can be effective in reducing pollution without dragging every polluter into court, and codification of generally accepted liability rules can lend orderliness and certainty to the common law process.

Between the common law and CWA regulations are water quality credit markets. Trading nutrient reduction credits does use governmental control to establish an acceptable level of pollution, but provides flexibility in achieving that level. At least this gets the incentives right on the pollution control side of the equation. Establishing markets in pollution credits, however, depends on possessing accurate data on the true state of polluted water bodies and the ability to quantify real reductions in pollution levels. Apart from such information, it is difficult-to-impossible to measure success or failure in restoring and maintaining water quality. Moving water quality markets forward will require improvements in measuring and monitoring pollution, especially from nonpoint sources, and institutional changes that improve the incentives for trade while improving the flexibility by which point and nonpoint sources can contract for improved water quality.

Well-defined, enforced, and tradable property rights give individuals an interest in monitoring their rights and finding ways to get additional environmental quality at minimum cost. Using incentives embedded in property rights and markets will take us further and more cost-effectively toward improving water quality than coercion has under the CWA.

9

THE RACE TO PUMP

As with many of our natural resources, concern is growing that groundwater supplies are being depleted by rapidly increasing demands. A 2009 article in *Scientific American* typifies this concern as it relates to agricultural production in America's Midwest:

> The Ogallala Aquifer, the vast underground reservoir that gives life to these fields, is disappearing. In some places, the groundwater is already gone. This is the breadbasket of America—the region that supplies at least one fifth of the total annual U.S. agricultural harvest. If the aquifer goes dry, more than $20 billion worth of food and fiber will vanish from the world's markets. And scientists say it will take natural processes 6,000 years to refill the reservoir.
> (Little 2009, 1)

The Ogallala, the largest in the High Plains aquifer system, is a particularly important groundwater resource in the United States. It underlies 174,000 square miles from South Dakota to Texas, and supplies water to approximately 27 percent of the irrigated land in the United States (Stewart 2003). In 2000 alone, the aquifer supplied 21 million acre-feet of water to 12.7 million acres of crops, up from 4 million acre-feet in 1949 (McGuire 2007, 1). Withdrawals from the Ogallala aquifer in 2000 accounted for 23 percent of the total withdrawals from all domestic aquifers for irrigation, public supply, and self-supplied industrial water uses combined; and 30 percent of the total withdrawals from all aquifers for irrigation (Maupin and Barber 2005, 1). The aquifer also provides drinking water to 82 percent of the population living above it (Dennehy 2000). In short, the Ogallala aquifer is a critically important water resource to the Midwest and, to a large extent, the nation.

Unfortunately, increasing demand on the Ogallala aquifer has outstripped the recharge rate, causing water levels to fall over time. Depth-to-water (distance from the surface to the top of the water table) ranges from 0 to 500 feet throughout the aquifer, with an average depth of roughly 100 feet. Since pumping intensified in the 1940s, water levels have declined more than 100 feet in parts of Kansas, New Mexico, Oklahoma, and Texas (Little 2009). Indeed, water level declines have made irrigation impossible or cost prohibitive in some areas overlying the Ogallala aquifer (Dennehy 2000). Total storage has also declined significantly. The United States Geological Survey (USGS) estimated that total water storage in the aquifer was about 2,925 million acre-feet in 2005, a decline of about 253 million acre-feet, or about 9 percent, since the 1950s (McGuire 2007).

These trends are not unique to the Ogallala aquifer. Signs of overdraft and declining groundwater storage are present in many of the nation's principal aquifers. In California's Central Valley aquifer system, for instance, the USGS estimates the average rate of decrease in storage between 1962 and 2003 at 1,900 cubic feet per second (Reilly et al. 2008, 48). For the Coastal Plain aquifer system along the southeastern seaboard, the estimated average rate of decline between 1999 and 2004 was 400 cubic feet per second (Reilly et al. 2008, 53). These are significant reductions in the amount of groundwater stored in these aquifers and, like the Ogallala aquifer, the explanation is quite simple: withdrawals exceed recharge. Much like a bank account with larger withdrawals than deposits, the balance of these groundwater resources is declining.

Nationwide, groundwater withdrawals have increased dramatically over the last half century. In 1955, total groundwater withdrawals in the United States were 47.6 billion gallons per day (BGD) (Kenny et al. 2009, 43). In 2005, that number had increased by 74 percent to 82.6 BGD. Though total groundwater withdrawals have been relatively flat since about 1975 (Kenny et al. 2009, 43), total groundwater storage continues to decline on a national scale because annual withdrawals exceed annual recharge for most aquifers.

Groundwater overdraft causes several environmental impacts. Perhaps the most prominent impact is land subsidence, and California's San Joaquin Valley provides a good example. Agricultural pumping in that valley has reduced groundwater resources by 60 million acre-feet and the water table by nearly 400 feet from pre-development levels (Weiser 2009). Combined with the area's collapsible soil structure, this intensity of pumping has caused vast areas of land subsidence. The most severe decline is near Mendota, where the ground has subsided 29 feet (Weiser 2009). In addition to infrastructure damage, land subsidence can permanently reduce the affected aquifer's storage capacity.

Another environmental impact of groundwater overdraft is salt water intrusion along coastal aquifers. Intense pumping can cause the saltwater interface to move inland via horizontal intrusion, over a drawn down aquifer's water table, or toward the pump intake in a process known as upconing, where saline water migrates upward toward the well through overlying freshwater (Delleur 2007). Florida's geology and coastal population distribution make that state particularly susceptible to saltwater intrusion of all types. In particular, South Florida's Biscayne aquifer has experienced significant saltwater intrusion following the recent increase in Miami–Dade county's population and groundwater demands. Moreover, the potential for sea level rise and increasing demands for freshwater in the region put the aquifer, and the region's water users, at considerable risk (South Florida Water Management District 2009).

The traditional response of the federal government to the problem of ground-water overdraft, land subsidence and saltwater intrusion has been to increase water supplies. For example, with the Water Resources Development Act (WRDA) of 1976,[1] Congress authorized $6 million

> to study the depletion of the natural resources of those regions of the States of Colorado, Kansas, New Mexico, Oklahoma, Texas, and Nebraska presently utilizing the declining water resources of the Ogallala Aquifer, and to develop plans to increase water supplies in the area and report thereon to Congress, together with any recommendations for further Congressional action.

The groundwater augmentation strategy has spanned several decades and continues today. In 2007, Congress passed the tenth WRDA since 1976. Like every WRDA that came before it, this legislation authorized millions of dollars of research into aquifer recharge and supply augmentation projects as a way to combat the environmental and economic consequences of intense groundwater pumping.

In contrast with the federal government, states and localities have responded to increasing groundwater depletion with demand side controls such as well spacing requirements, limits on groundwater pumping, and closure of basins to new wells. One of the most common regulations in groundwater law has been well spacing limitations that establish a minimum distance between wells. Several western states have passed legislation designating critical groundwater areas where existing pumping may be restricted and new pumping may be prohibited to preserve a sustainable amount of groundwater.

As with surface water, groundwater crises are related to the institutional and legal frameworks governing the resource. Without the proper information and incentives, private users are not likely to augment groundwater supplies or reduce their demands on groundwater basins. The institutions that currently provide the information and incentives are a combination of often vague property rights and central, bureaucratic agencies that dictate water allocation. When groundwater was abundant, the nature of these institutions made little difference; but growing demand has placed the claims on groundwater in direct competition with one another. Problems with drawdown, land subsidence, saltwater intrusion, rising pumping costs and pollution suggest the need for institutional reform that pays attention to the structure of property rights that provides incentives for conservation and efficient allocation.

A Primer on Hydrogeology

Before discussing how groundwater institutions might affect allocation, it is important to understand the basic physical characteristics of the resource. For the most part, groundwater does not flow in streams beneath the earth's surface, nor do groundwater basins resemble lakes. The hydrology of groundwater is not so simple. Groundwater is found in areas where porous materials, such as sand, clay or rock underlie the surface. Water percolates downward from the earth's surface until it reaches an impermeable layer. The geological formations in which groundwater is stored are the aquifers. A simple model of an aquifer would be a bowl filled with sand. Water poured onto the sand would percolate downward to the bottom of the bowl, eventually saturating the sand and forming a pool. The top of this pool is known as the water table.

Aquifers differ in shape and size and in their ability to store water. These differences depend on a multitude of hydrological factors. For example, the rate at which water moves through an aquifer depends on the permeability or hydraulic conductivity of the rock, and the storage capacity depends on porosity.[2] The combination of permeability and porosity determines whether aquifers are capable of storing and supplying recoverable quantities of water (Heath 1988, 76–77).

Discharge from an aquifer consists of seepage from the aquifer into surface streams and evapotranspiration through plants whose roots extend into the water table. Recharge occurs via seepage from streams, lakes and land areas where rainfall and runoff percolate through the ground into the aquifer. Because of the interconnections between groundwater and surface water, the impacts of pumping from an aquifer on surface water users must be taken into account (Sax and Abrams 1986, 837).

"Under natural conditions, prior to the development of wells, a groundwater system exists in a state of approximate equilibrium" (Glennon and Maddock 1994, 577), meaning discharge and recharge are in balance. When pumping begins, the equilibrium is disrupted and two things happen. First, the water initially pumped comes from water stored in the aquifer. Second, pumping can eventually draw additional recharge from any connected surface water or from nearby aquifers. This transition from storage to increased recharge may vary from days to millennia, depending on the characteristics of the aquifer. Where the transition is slow or connected supplies are absent, pumping will decrease the amount of water stored in the aquifer. Where the transition from groundwater storage to increased recharge is fast, surface water supplies will decline in response to groundwater depletion (Balleau 1988). Either way, if groundwater withdrawals exceed natural recharge and induced recharge, overdraft will result, and the aquifer's water level will drop.

The amount of water stored in an aquifer, known as groundwater stocks, serves two important purposes. First, stocks determine pumping lift costs, and second, they provide a source of insurance against changes in precipitation or surface-water availability. Hence, groundwater stocks can be drawn down or augmented depending on expectations of recharge.

If there is open access to pumping from the stocks, a race to the pump house where groundwater is pumped early and fast occurs for three reasons. First, a tragedy of the commons occurs because of the rule of capture. In this setting, groundwater not captured now by one user will be used by others. Hence there is little incentive to conserve for the future. Second, not only is conserved water unlikely to be available in the future, it will cost more to pump it if the water table declines as a result of overdraft. To avoid higher pumping costs, again users have an incentive to pump now rather than later. Finally, where the aquifer is permeable and lateral movement of water is rapid, pumping at one location can have a direct impact on the water table and water pressure at nearby locations. This happens because pumping can cause the water table around a well to drop, in turn creating a pressure gradient between the well and surrounding aquifer. As water moves from higher water levels or pressures toward lower levels or pressures surrounding the well, it's pulled from the surrounding aquifer and any nearby wells. This in turn can reduce water levels around other users and can even cause other wells to run dry.

Various aquifer characteristics and groundwater use either aggravate or mitigate well interference among pumpers. An impervious aquifer, for example, limits the transmission of groundwater from one location to another, limiting the possibility of well interference.[3] Also, if groundwater is costly to pump and transport, its use might be limited to lands overlying the aquifer. Hence, diminishing returns from applying groundwater to a fixed land base can constrain pumping. An expanding agricultural land base or increasing municipal and industrial uses, however, can offset diminishing returns and increase withdrawals. Other aquifer characteristics such as depth, purity, rate of recharge, pressure at which the water is stored, and connection to other aquifers and surface waters will affect well location, pumping costs, well interference, and amounts and timing of pumping.

In short, hydrogeology makes the problem of defining and enforcing private groundwater rights more complicated than for other resources. Water stored underground is more difficult to measure, hydrological connections between surface and groundwater are not always obvious nor easily assessed, and third-party impacts associated with drawdown can be significant.

On the other hand, the science of groundwater hydrology has become more precise in recent years and continues to improve. These advancements in understanding make the definition and enforcement of rights to groundwater easier. Moreover, as pumping from groundwater basins increases, additional data become available to help determine recharge, discharge, connectivity with surface waters, and permeability (transmissivity), making it easier to specify groundwater rights. As with any property rights to natural resources, institutional evolution occurs as the groundwater becomes scarcer and inefficient allocation becomes more costly.

Groundwater Rights Evolving Toward Markets

The evolution of groundwater rights has followed a pattern similar to that of surface water rights. When water was abundant, little investment was needed in developing institutions that would ensure efficient allocation and use decisions. As long as groundwater was not scarce, it made little sense for the early American settlers to devote much effort to devising market-based institutions to govern its allocation.

As groundwater stocks and water quality concerns began to surface, the need for institutional and legal changes became clear. In response, groundwater rights have, over time, become better defined, enforced, and divestible.

Absolute Ownership

As with many of our property institutions, English common law rules were the simplest to adopt; and the English rule of absolute ownership or the rule of capture was first used to establish property rights in groundwater. That rule gave the overlying landowner complete freedom to extract groundwater from under his land and use it on or off his property without liability. Precedent for the rule was established in the English case of *Acton v. Blundell*[4] in 1843, where Lord Chief Justice Tindal gave the following opinion:

> [I]n the case of a well sunk by a propriator [sic] in his own land, the water which feeds it from a neighbouring soil does not flow openly in the sight of the neighbouring propriator [sic], but through the veins of the earth beneath its surface; no man can tell what changes these underground sources have undergone in the progress of time. It may well be, that it is only yesterday's data, that they first took the course and direction which enabled them to supply the well: again, no propriator [sic] knows what portion of water is taken from beneath his own soil: how much he gives originally, or how much he transmits only, or how much he receives . . . [5]

Because the hydrology of groundwater was unknown to early English courts, they avoided the issue by classifying groundwater as property with the same status as rocks or minerals on or under the land. According to Frank Trelease (1976, 272), it "was in the light of this scientific and judicial ignorance that the overlying landowner was given total dominion over his 'property,' that is, a free hand to do as he pleased with water he found within his land, without accountability for damage." This form of property rights worked well as long as third-party injuries were rare; that is, as long as groundwater was not scarce.

Reasonable Use Doctrine

As the demand for water grew in the United States and individuals began to compete for water and land use, the English rule of absolute ownership had to be modified. United States courts softened the English rule with the reasonable use rule, a modification of the absolute ownership rule that required groundwater use have a reasonable relationship to the use of the overlying land. The distinguishing feature of the reasonable use rule was its prohibition of wasteful and off-site uses, both ostensibly permissible uses under the rule of absolute ownership.

Though the reasonable use rule did curtail some groundwater uses, the name is somewhat misleading because, like the absolute ownership rule, the reasonable use rule of groundwater allocation imposed no liability on or injunctive relief against a user who negatively impacts the groundwater supplies of neighboring landowners.

As the doctrine has developed, it generally has been held that

> all uses of water upon the land from which it is extracted are "reasonable," even if they more or less deplete the supply to the harm of neighbors, unless the purpose is malicious or the water simply wasted. But . . . when the question is whether water may be transported off that land for use elsewhere, this is usually found "unreasonable", though it has sometimes been permitted. Authorities are not all agreed, but a principle that seems to harmonize the decisions is that water may be extracted for use elsewhere only up to the point that it begins to injure owners within the aquifer.
>
> (Stoebuck and Whitman 2000, §7.5, 428–29)

Indeed, the original doctrine of reasonable use did not consider the reasonableness of a groundwater use beyond the location of the use and the intent of the user. "Reasonableness" in the groundwater context thus differed from the reasonableness standard embodied in the riparian doctrine, the latter describing a correlative right between water users.

Modified Reasonable Use and Groundwater Permits

During the latter half of the twentieth century, several state courts and legislatures modified the reasonable use rule to address injury to neighboring groundwater users. For example, in Illinois, the Water Use Act of 1983 defined "reasonable use" as "the use of water to meet natural wants and a fair share for artificial wants."[6] Similarly, Georgia requires pumpers provide documented evidence that the amount of water proposed to be withdrawn or used is "reasonably necessary to meet the needs of the applicant" and that such withdrawals will not impose "unreasonable adverse effects on other water uses or users . . . including public use."[7] Such modifications made the reasonable use rule of groundwater allocation correlative in nature and thus consistent with the riparian doctrine for surface waters. However, this consistency in the law came at the cost of increased uncertainty over groundwater rights given the subjectivity of such concepts as "reasonable use," "fair share," and "adverse effect."

The Restatement (Second) of Torts Section 858 (1979)[8] created additional uncertainty over groundwater rights by allowing courts to balance the reasonableness of a withdrawal against the nature and extent of the harm caused to other groundwater users. In calling for courts to balance the equities and hardships between competing groundwater users, the restatement encourages courts to ignore previously established use rights and consider the totality of circumstances of a given groundwater dispute. This approach gives courts carte blanche to examine the relative wealth of the parties to determine who is better able to bear the burden of the harm caused by the withdrawal. The goal is an overall determination of "fairness" in each case. Fortunately, only a minority of states has adopted the restatement approach.

In conjunction with this modification of the reasonable use rule, several eastern states adopted groundwater permitting processes that incorporate the principles of either the original or modified reasonable use doctrine. Though each state's groundwater permit system is unique, several important variations include: the criteria for permit issuance or denial, the minimum withdrawal amount requiring a permit, permit transferability, and the hierarchy of uses during times of shortage.[9] Permits may apply to groundwater exclusively or to both surface and groundwater withdrawals. They may also apply statewide or to designated districts.

In the specified location and for the specified type of withdrawal, these permits are the exclusive basis for non-exempt withdrawals. Nonetheless, groundwater permits constitute only usufructuary rights which are revocable without compensation. They do not create private ownership of groundwater resources or a legally recognized property right. Consequently, they create no additional incentive for individual pumpers to conserve groundwater resources.

Prior Appropriation Doctrine

In western states, groundwater conflicts could not be avoided as populations grew. As with surface water, increasing demands for groundwater led to the repudiation of the reasonable use doctrine in favor of allocating rights to groundwater based on prior appropriation. Rather than tying water rights to landownership, however, appropriative groundwater rights (like surface water rights) are acquired by application of the water to a beneficial use on a first-in-time, first-in-right basis. In times of shortage, junior rights holders may be prevented from pumping until senior rights are satisfied.

Most western states now require a permit to withdraw groundwater, with the standards for granting the permits roughly equivalent to those that must be met in order to acquire an appropriative right for surface water use. In Idaho, for example, new groundwater users must follow the same application process as new surface water users, including obtaining a permit from the state, creating a physical diversion (well), and putting the water to a beneficial use as defined by the state. As with surface water rights under the prior appropriation doctrine, permitted groundwater withdrawals are not limited to use on overlying lands. And any change of use or transfer (where permitted), must be approved by the appropriate state agency. The conditions of transferability vary by state, but the general rule is that any change in groundwater use caused by a transfer of right must not harm other vested rights.

California Groundwater Rights

California has adopted a combination of groundwater management doctrines resulting in three distinct categories of private groundwater rights. The first category of groundwater rights is overlying rights. These exist by virtue of landownership above a groundwater aquifer, and allow the overlying landowners to pump water from the basin for reasonable, beneficial purposes on the land. Overlying rights are correlative in nature, meaning overlying landowners must reduce groundwater use on a pro rata basis during times when groundwater demands exceed supplies.[10] The proration is based either on the extent of overlying property or amount of groundwater previously withdrawn. Such an overlying right conveys with the land regardless of whether the right is currently or has ever been exercised.

Appropriative rights are the second category of groundwater rights in California. These rights originate through the reasonable and beneficial use of groundwater in excess of the reasonable needs of overlying landowners. Between groundwater appropriators, such surplus waters are apportioned under the doctrine of first in time, first in right. Between overlying and appropriative rights, however, overlying rights are superior absent claims of prescription.[11]

Prescriptive rights constitute the third category of California groundwater rights and only come into existence when there is no surplus water available in a groundwater basin.[12] Prescriptive rights were first recognized by the California Supreme Court in *City of Pasadena v. Alhambra*.[13] The Pasadena case involved the adjudication of the Raymond basin, which had been overdrafted for years by both overlying correlative rights holders and non-overlying prior appropriators. Rather than shutting down the non-overlying prior appropriators and reducing the correlative rights pumpers on a pro rata basis, the court held that the prior appropriators had acquired prescriptive rights equal to the correlative rights in the basin. The Pasadena court held that once overdraft had begun, each pumper was acting prescriptively to other pumpers in the basin.[14] After determining the basin's safe yield, the Pasadena court reduced all pumping on a pro rata basis based on amounts pumped by each right holder in the 5 years prior to initiation of the case.

By clarifying both the correlative and appropriative rights in the basin, the Pasadena case minimized the common pool problem, reduced tenure uncertainty, and facilitated groundwater transfers (Garner et al. 1994, 1025–29). Unfortunately, the case also created a race to the pump house incentive for both prior appropriators and overlying landowners in other groundwater basins in California. In ordering a pro rata reduction of all pumpers, the court encouraged groundwater users to establish as large a present use as possible so that any pro rata reduction would be figured in relation to a larger initial quantity of water (Sax and Abrams 1986, 830).

Several cases have since limited the application of the prescription doctrine. In the 1975 case of *City of Los Angeles v. City of San Fernando*,[15] the California Supreme Court rejected prescriptive right claims against rights held by the city of Los Angeles. Basing its ruling on section 1007 of the California Water Code, the California Supreme Court held municipalities were immune to prescription of their groundwater rights. More recently, the California Supreme Court confirmed the importance of the traditional priority system by rejecting the concept of "equitable apportionment," a prescription-like argument that courts should allocate water in overdrafted basins based on notions of fairness rather than property rights.[16] These decisions have helped clarify the hierarchy of groundwater rights in California, though the topic remains contentious and frequently litigated.

Conjunctive Water Management

Nationwide, few states govern water in a way that fully accounts for the hydrological connection between ground and surface water resources. Instead, most states operate separate institutional systems for surface water rights and groundwater rights. Although several states include in their surface water rights system any groundwater located directly beneath the cut bank of the stream, this distinction is hydrologically arbitrary because it fails to account for the fact that groundwater outside the cut bank of the stream may also be connected to the stream. The difficulty in determining the connection between ground and surface waters has led many states to ignore this fundamental connection in their water management structures. "A complete misunderstanding of hydrology has been memorialized in many states, where groundwater and surface water are legally two unrelated things" (Glennon 2002, 210).

Montana is one exception. In 2007, the Montana legislature passed House Bill 831 which requires a hydrogeologic assessment for all new groundwater permit applications in closed basins. If the assessment predicts a statutorily defined "adverse effect" on any prior appropriator of surface or groundwater, the applicant must offset any groundwater depletions with a mitigation or aquifer recharge plan. This policy recognizes the hydrological connection between surface and groundwater and embodies a rights-based approach to resolving conflicts over that connection.[17]

The groundwater doctrines existing today have evolved as a result of changes in the benefits and costs of defining and enforcing property rights. The arid western states follow the prior appropriation doctrine, with California using the correlative rights doctrine as well. The more humid eastern states allocate groundwater according to varying principles of reasonableness. The differences between the East and West can be traced to the higher annual precipitation experienced in the East that has provided little incentive to modify the original doctrines of absolute ownership and reasonable use. It is not surprising that as some eastern states experienced groundwater overdraft, they have begun to modify their laws to conform more closely to available supplies.

While it is true that the changing relative scarcity of groundwater has brought pressure to change groundwater institutions, the result has not necessarily been well-defined and enforced property rights that encourage efficient use. Current institutions are deficient both in terms of tenure certainty and transferability. As long as rights to groundwater are poorly defined, unenforced, or unable to be traded to higher valued uses, water will not be conserved and used efficiently. Under the rule of capture, water not pumped today might not be there tomorrow. In reasonable use states, limits on the transferability of groundwater rights to non-overlying lands reduce efficiency by prohibiting water allocation to higher valued uses. In prior appropriation states, the use-it-or-lose-it principle similarly discourages conservation for future use. The increasing scarcity of groundwater resources is forcing states to consider alternative institutions, including markets.

Creating Private Rights to Groundwater

The big hurdle in the evolution of groundwater rights is resistance to changing the status quo use of groundwater and to changing the bureaucratic mechanisms that determine that status quo use. Individuals who are now pumping from an aquifer are not likely to voluntarily reduce their rate of extraction. Nor are bureaucratic agencies likely to give up power they may have over allocation. Getting all parties to buy into privatization requires demonstrating how market forces can enlarge the pie through improved allocation efficiency and determining which government institutions are necessary to facilitate this improvement.

Privatizing groundwater basins can give pumpers an incentive to optimize groundwater extraction. If a basin were owned by a single individual, that person would extract at a rate that would maximize the basin's net present value, taking into account the current and future value of groundwater stocks and any connections between pumps. For the individual owner, the immediate value of stocks would be a function of the value of water in production or consumption. The future value will depend on how changes in the stocks impact pumping costs, spatial distribution, and insurance value. Specifically, larger stocks will have a positive value to the extent that they reduce pumping costs, expand the surface area of the aquifer, or provide an insurance value against annual variations in the availability of surface water. As the water table declines, pumping costs rise, the aquifer's surface area decreases, and less water is available for future use.

Stock and Flow Rights

Though a single owner might take all of these values into account, multiple pumpers acting independently will only do so if rights to the groundwater stocks and flows are well defined and enforced and if third-party effects are minimized. In 1977, Nobel laureate Vernon Smith proposed a way of doing this. Smith's scheme would issue a property deed for a share of groundwater from a particular basin to each individual groundwater user I, $I = 1, 2 \ldots n$, for n users in total. Each deed would have two components, one allowing claim to a percentage of the annual recharge or flow into the basin, and the other to a percentage of the basin's storage or stock. The property rights would be allocated to individuals in proportion to their pumping rates during a specific base time period.

To illustrate how the concept would work, Smith used the Tucson Basin and 1975 as the base period. Based on that year's extraction of 224,600 acre-feet, each individual proportion of stocks and flows, P_i, would be $x I / 224,600$, where $x I$ is the amount of water used by the individual in 1975. Based on that proportion, P_i, each pumper would receive two rights:

1. The flow right would be based on a fraction of a long-run average of the net recharge to the basin, which for the Tucson Basin was estimated to be 74.6 thousand acre-feet. Therefore, the property right of individual I to the annual recharge would 74.6 P_i thousand acre-feet in perpetuity.
2. The second property deed would convey a right to a share of the basin's stock, which was approximately 30 million acre-feet in 1975. The share of this stock granted to individual I would also be P_i.

(Smith 1977, 7–10)

The initial allocation of water rights can promote waste, especially if it induces a race to the pump house (see Anderson and Hill 1981). For example, if policymakers were considering the idea of privatizing a groundwater basin based on average use during the 5 years prior to privatization, a race to the pump house could ensue among pumpers wanting to increase their share of rights. To avoid this race to increase individual rights, water deeds could be based on use over a longer period of time or they could be assigned to landowners in proportion to ownership of land overlying the aquifer.

Tracking Groundwater Use in a Privatized Basin

Given some initial allocation of rights to stocks and flows such as that proposed by Smith, enforcement of rights requires an operational accounting system. Pumps would be metered, and each owner of a right would begin with an initial stock. Maintaining meter integrity would be similar to that of existing power and water meters. At the end of each year, an adjustment would be made to the stock account by subtracting the amount pumped and adding the appropriate share of aggregate recharge. Because recharge amounts are subject to a host of hydrologic variables, a running 5-year average could be used and updated annually. If return flows from irrigation were considerable, they could be added to an individual's stock or recharge account. In either case, return flows should be tracked and attributed to individual users to maintain an incentive to reduce the consumptive use of each share of the groundwater resource. Violations for pumping water in excess of the amount owned would be handled with fines or deducted with a penalty from the pumper's share of subsequent years' recharge.

This type of accounting system has been in place since 1978 in the Genèvois Basin, which underlies the border between France and Switzerland. A treaty signed by the Canton of Geneva and the Prefecture of Haute-Savoie created a commission to supervise groundwater use.

> The commission keeps a complete inventory of public and private pumping installations in the two countries. Each installation has a metering device indicating the volume of water taken by each user . . . France's contribution to defraying the recharge costs is assessed by reference to the amount of water taken by French users together with the contribution to the natural recharge

of the aquifer made by French territory . . . The commission has at its disposal a system of control which allows it to know with certainty the intensity of use of the aquifer and thus plan withdrawals rationally with the needs of users in mind.

(Barberis 1991, 185)

Groundwater Transfers

Allowing transfers among owners in a groundwater basin improves allocation in at least two ways. First, voluntary transfers allow water to move from lower to higher valued uses. Second, transfers allow individuals to better deal with risk relating to variations in water availability. Contracting among water owners offers an opportunity for the relatively more risk-averse parties to trade with the relatively less risk-averse parties. Those who are relatively risk-averse will want to hold stocks as a contingency against water shortage associated with random fluctuations in recharge and demand. As a result of these market transactions, prices of groundwater rights will reflect differential risk preferences among the producers in the basin.[18]

Impacts on right holders not involved in groundwater transfers, often called third-party impacts, drive the principle objection to groundwater privatization and marketing. Thus, any system for privatizing groundwater and allowing transfers between willing buyers and sellers must protect the groundwater rights of third parties. Consider the case in which an individual with a right to extract groundwater from an aquifer sells the water to someone outside the basin or to someone who changes the use in such a way that it reduces the return flow or recharge to the aquifer. Third-party rights holders could face increased pumping costs or, worse, they could be unable to extract the full amount of groundwater to which they have a right.

The most straightforward way to protect third-party rights from infringement of this sort is to restrict groundwater transfers or changes of use to the consumptively used portion of the right. For example, if consumptive use were measured and determined to average 65 percent of each water right in an aquifer, a person who wanted to transfer 130 acre-feet out of the basin would have to relinquish his right to 200 acre-feet. The remaining 70 acre-feet (35 percent) would be left in the aquifer to provide what would have been recharged.

While the exchange of consumptive rights to stocks and flows improves efficiency with respect to use and allows individuals to share risk, the property rights system does not completely solve the commonality problems associated with pumping costs and spatial distribution. An individual user may hold title to a certain stock of water, but the conditions under which he can obtain the water in the future are altered by the use rates of other pumpers in the basin. First, rapid pumping in the aggregate will increase the future cost of pumping for any individual. Second, an individual with a well near the perimeter of the aquifer who conserves his stock could find his well dry if the aquifer is drawn down by other pumpers. The stock rights could be sold but the value would depreciate with a declining surface area of land overlying the aquifer.

To achieve efficiency, given positive pumping costs and the potential that some overlying lands might be left dry by basin drawdown, Smith's property rights scheme requires slight modification. When initial stocks are allocated to private owners, some stocks must be withheld from allocation to ensure that the water level in the basin does not fall too low. The ideal or equilibrium level of stocks to be withheld, however, is difficult to determine because many of the hydrologic and economic variables are random and subjective. Improved information will be available after the initial appropriation of stocks for private property has been made and, ex post, better estimates will evolve as pumping occurs. Unless there are significant economic or hydrologic surprises, the economic costs of error and the choice of stocks withheld should be relatively small.

Unitization

Unitization is an alternative means of achieving efficiency in the face of pumping costs and well interference. It is an approach that has been used for oil and natural gas deposits that lie in subterranean reservoirs similar to groundwater. When a basin is unitized, it is managed by a cooperative agreement among the overlying landowners or their lessees. Thus, unitization provides an opportunity to manage a groundwater basin as if it were owned by a single entity. When individual pumpers unitize a basin, they agree to develop it as a whole and divide the costs and profits proportionally. This arrangement enables them to obtain the most efficient production from the field by carefully spacing wells and by collectively determining the reservoir-specific rate of extraction. By drilling the optimal number of wells in strategic locations over the aquifer, developers can minimize pumping costs and optimize the rate of extraction. In addition, extraction can be controlled and adjusted in response to market conditions. Monitoring and enforcement costs are reduced because the resource is recovered from a few, closely controlled wells. In short, unitization "eliminates competitive withdrawal and directs extraction toward maximization of the economic value of the entire reservoir, rather than of the segments (leases) held by individual parties" (Libecap 2008, 224).

Faced with inefficiencies caused by excessive pumping under the rule of capture, some oil and natural gas producing states have required unitization—with favorable outcomes. As a result of a compulsory unitization statute for oil and gas reservoirs, Louisiana's oil and gas wells are, on average, one-third more productive than those in Texas, which does not require unitization (Murray and Cross 1992, 1150).

Although states have not formally adopted unitization policies for groundwater resources, informal agreements are being tested on a basin-level scale. In Utah's Escalante Valley, decades of water-intensive farming and groundwater allocations far in excess of recharge rates have caused groundwater levels to drop more than 100 feet since 1948, leading to land subsistence and declines in water quality (Hansen 2010). The state, in 2007, proposed a groundwater management plan calling for reductions in groundwater use by nearly half over 90 years. The state's plan would retire certain water rights while placing withdrawal restrictions on remaining right holders. The management plan also contained a provision that would provide for voluntary arrangements among right holders to return groundwater levels to sustainable levels.

Groundwater users quickly took advantage of this opportunity and countered with their own plan. Similar to unitization, they proposed a plan that would pool water rights within the basin, then collectively make the necessary reductions in withdrawals. In addition they introduced a bill (SB-20) that, if passed, would create a community-run water district authorized to manage the proposed groundwater management plan (Escalante Valley Water Users Association 2011). The bill would effectively transfer management decisions from the state engineer to valley residents. With an increasing number of groundwater basins facing similar circumstances and challenges going forward, the unitization experiment in Escalante valley may prove to be model for other basins.

Unitization, coupled with quantified, transferable groundwater rights, could allow for more efficient development of groundwater aquifers. By drilling and utilizing an optimal number of wells, pumping costs as well as monitoring and enforcement costs could be reduced. The collection of pumpers overlying a unitized aquifer could internally manage well interference problems. Although the reduction in the number of wells would require transportation of pumped water to land on which it is used, that cost must be weighed against the prospect of third-party effects and peripheral wells abandoned by basin drawdown. Moreover, these higher costs might be minimized by the existing surface water infrastructure of irrigation districts, mutual ditch companies, and federal water projects.

Of course, the transaction costs of cooperative agreements for development of an aquifer may be so high as to outweigh the benefits of unitization in some basins, particularly where the number of existing pumpers is large. But as groundwater becomes scarcer, the costs of inefficient allocation rise, thus increasing the incentive to overcome those transaction costs.

Assigning private rights should induce less extraction because the groundwater will have tangible value to individual users, which is not the case under the rule of capture or reasonable use doctrines. As producers compare water table declines under a private ownership regime with experiences of uncontrolled pumping, they are likely to observe more conservative behavior. This should generate expectations that pumping costs will increase less rapidly in the future than they would in a commons setting. As long as pumpers adjust to this expectation, rates of withdrawal over time should approach optimality.

Because groundwater markets require definition and enforcement of rights to stocks and flows and because this definition and enforcement will not eliminate all third-party impacts, many responsibilities remain for government agencies. A private property rights regime would place allocation decisions in private hands, generate price information about values in different uses, and encourage mutually beneficial trades. But all of this would depend upon centralized recording of titles and protection of third parties. As with surface water rights in prior appropriation states, state water agencies would have responsibility for determining levels of existing stocks, recharge rates, and consumptive use rates. To guard against pumping and spatial distribution effects (in the absence of a private unitization agreement), at minimum, agencies would have to provide a hearing process where third parties could challenge market transfers, and at a maximum would have to determine allowable pumping and interference rates.

Private Rights in Practice

Assigning private, transferable rights to groundwater can improve allocative efficiency, but do the potential gains justify the costs of establishing those rights? The answer is no if groundwater is relatively abundant and the costs of defining and enforcing rights exceed the benefits. But as groundwater scarcity threatens pumpers with increased costs and limited supplies, market-based allocation schemes become more attractive and even necessary, as the following examples illustrate.

Tehachapi Basin, California

In California, where groundwater scarcity is common, adjudication has become the primary mechanism for privatizing groundwater. Though adjudication is time- and resource-intensive, the resulting clarity and transferability of groundwater rights promote allocation efficiency.

Because California does not have a comprehensive groundwater management statute or program, local public entities are primarily responsible for groundwater management. In unadjudicated basins, a variety of public entities such as cities, counties, groundwater management districts, water user agencies, or any combination thereof may share in groundwater management responsibilities. These entities may impose pumping controls or restrictions on export to reduce basin overdraft. In adjudicated basins, however, court-appointed watermasters have management authority and may curtail water use by more junior right holders in times of reduced supplies. Rights in these basins are quantified and transferable, setting the stage for groundwater markets.

The Tehachapi Basin is located in Kern County, California, approximately 35 miles southwest of Bakersfield and 100 miles north of Los Angeles. The basin was adjudicated in 1973. The 37-square-mile basin is the largest of three adjudicated basins in the area covered by the Tehachapi–Cummings County Water District. Water in the basin is used primarily for agriculture, but municipal and industrial uses have increased substantially in the past few decades. The only source of natural recharge to the basin is precipitation in the watershed.

Groundwater overdraft began in the Tehachapi Basin in the 1930s following a steady increase in irrigation.[19] By 1960, withdrawals exceeded recharge by 60 percent. The water level basin-wide dropped by an average of 70 feet between 1951 and 1961, while the level around the city of Tehachapi fell 110 feet. During that time, groundwater storage in the valley fell by over 61,000 acre-feet. From 1961 to 1968, the water table continued to drop an average of 3 feet per year.

Consequently, pumping costs increased substantially, and some wells ran dry. Fears that continued overdraft would seriously affect the agriculturally based economy brought about the formation of the Tehachapi–Cummings County Water District.

In 1965, a citizen advisory committee was formed to consider the options for managing the basin. The committee decided to bring in surface water from the California Aqueduct and to adjudicate groundwater water rights in the basin. Because the basin is situated at an elevation of 4,000 feet, importing surface water required a pipeline to lift aqueduct water over 3,400 feet. That made surface water far more costly than groundwater and that, in turn, made it unlikely that users would find substitutes for groundwater without some additional incentive.

Adjudication of the Tehachapi Basin offered the main hope for controlling overdraft. In 1966, the water district filed suit in Kern County Superior Court, asking the court to adjudicate groundwater rights in the basin. The judgment, handed down 5 years later, followed the mutual prescription doctrine and limited total extraction to the basin's hydrologically determined safe yield. The court quantified each party's base groundwater rights equal to the highest average annual extraction rate over any consecutive 5-year period after basin overdraft began in the 1930s. The base rights totaled 8,250 acre-feet for the basin, while safe yield was estimated to be two-thirds of the base rights, or 5,500 acre-feet. The court thus allocated to each party two-thirds of the base right. The court also ruled that users pumping less than their allocated amount could stockpile part of the excess for up to 2 years, but the amount stockpiled was limited to 25 percent of the allowed allocation.[20] The costs of adjudication totaled $300,000 for 100 users, or less than $55 per acre-foot.

To encourage the use of imported surface water that became available in late 1973, an exchange pool was established. The pool allowed reimbursement of users located near the surface water for the difference between surface water costs and average groundwater pumping costs. Suppose, for example, that surface water for agricultural use was priced at $100 per acre-foot, and average groundwater pumping costs were $40 per acre-foot.[21] If an individual not adjacent to the surface water source wanted to use more than his allocated share of groundwater, the watermaster could allow him to pump groundwater in excess of the adjudicated right at a charge of $60 per acre-foot. During the same period, a user adjoining the surface water source would be required to substitute an equivalent amount of surface water for groundwater. That user would be reimbursed for the $60 per acre-foot difference between the price of surface water and the average groundwater pumping costs.

In addition to the exchange pool, which allows users to substitute groundwater for imported surface water, Tehachapi Basin pumpers directly trade groundwater pumping rights using short-term leases and permanent transfers. For instance, Stallion Springs and Bear Valley Springs community service districts, because of high pumping costs in their location, routinely purchase and pump water from rights holders located lower in the basin. In 2008 alone, short-term transfers totaled 2,366 acre-feet, or roughly 43 percent of the pumping rights allocated under the 1973 adjudication (Tehachapi–Cummings County Water District 2008, 10).

There are, however, two restrictions on the exchange of groundwater rights in the Tehachapi Basin. First, the Kern County assessor has ruled that water rights severed from the land are subject to the same taxes as mineral rights. Because these taxes are high, the exchange of groundwater has occurred only by sale of the overlying land or short-term leases. Second, to guard against third-party impairment, the watermaster must approve extraction at a location other than where the water right was developed. When a water right is leased or sold to another party, the water is not physically transported; it is pumped from the lessee's or buyer's well that may be in another part of the basin. The rationale for requiring watermaster approval is that a substantial number of transfers could create cones of depression in the area where pumping is increased and thus adversely affect other pumpers.

Two lessons can be learned from the adjudication of the Tehachapi Basin. First, rights to groundwater can be defined and enforced. While the Tehachapi adjudication involved only rights to recharge, the fact that water can be stockpiled suggests that definition and enforcement of rights to the groundwater stock are also feasible. Second, observations show that major externalities in the Tehachapi Basin have been addressed through adjudication. During the period between early development in the 1950s through adjudication in the 1970s, groundwater levels declined steadily, falling roughly 100 feet below pre-development levels. Following adjudication and changes in groundwater management, levels have since recovered to levels similar to the 1950s (California Department of Water Resources 2006). Additionally, the importation of surface water has provided incidental recharge and that in turn has increased the water table in some areas. The city of Tehachapi no longer rations water as it did during some periods prior to adjudication, and rising water tables have brought previously marginal wells back into production.

Though limited privatization of rights to recharge has produced some improvements, further decentralization and privatization could increase efficiency even more. In particular, eliminating the 25 percent cap on groundwater stockpiling and defining stock rights—in addition to flow rights—would allow rights holders to better hedge the risk of water scarcity and to fully utilize the basin's resources. One state that allows for the full utilization of its groundwater resources is Oklahoma.

Oklahoma

The evolution of Oklahoma's groundwater law is an illustrative example of how legal institutions change in response to water scarcity. The first rule applied to groundwater management in Oklahoma was the English common law rule of absolute ownership. As stated earlier in the chapter, absolute ownership gives each landowner complete freedom to pump as much groundwater as he needs, without liability for the impact of that pumping on his neighbors. Given the abundant groundwater resources underlying Oklahoma, conflict between pumpers was relatively seldom and, consequently, absolute ownership remained the rule until 1936.

In that year, the Oklahoma Supreme Court decided *Canada v. Shawnee*[22] and shifted state policy to the American rule of reasonable use. The move reflected the growing concern over water scarcity and the increasing frequency of groundwater conflicts. Following this decision, pumpers were required to make reasonable use of the groundwater they pumped, a limitation which excluded the transfer of water off-site. Though the shift to reasonable use might have alleviated Oklahoma's race to the pump house, it left water users no mechanism for reallocating water to higher valued uses.

As demands for new water sources continued to grow and groundwater development expanded, the reasonable use standard soon proved ineffective in addressing growing conflict and scarcity. In 1949, the Oklahoma legislature enacted a permit-based appropriation system for both surface and groundwater in a move that opened the door for groundwater marketing. The legislation called for court adjudication of groundwater pumping rights based on hydrologic surveys of each basin's safe yield. Safe yield was limited to natural recharge in an effort to prevent overdraft. By limiting pumping rights to groundwater flows and by allocating rights via court adjudication, Oklahoma's 1949 groundwater law closely resembled California's current allocation system. However, the law was never implemented as it was written because the basin safe yield limitation entailed economic tradeoffs unsuitable to both pumpers and politicians.[23]

In 1973, the Oklahoma legislature rewrote the state's groundwater law to embrace a policy of utilization over preservation (Roberts and Gros 1987, 538). Similar to the 1949 law, the 1973 law required hydrologic surveys and safe yield determinations for each basin. The critical difference was that safe yield under the 1973 law was determined under the assumption of a 20-year basin life span. Groundwater pumping rights were then proportionally allocated based on landownership. Each overlying landowner received the same per-acre allocation, except for landowners with prior appropriation rights who were protected from any reduction in their rights.[24]

In practical effect, this allowance of basin depletion over a fixed term constitutes an allocation of rights to both groundwater flows and stocks. Combined with the simple apportionment of rights on a per-acre basis—as opposed to lengthy courtroom adjudication—Oklahoma's 1973 groundwater law ensures full utilization of the resource and reduced administrative inefficiencies.

Before rights can be assigned, however, hydrologic studies must be completed to determine the amount of water in each basin and the basin's maximum annual yield. Such studies are costly and sometimes take years to complete. Nearly 40 years after passage of the 1973 legislation, hydrologic investigations and maximum annual yield determinations have been completed on 16 of the state's 23 major basins, and only seven of the estimated 150 minor basins (Oklahoma Department of Environmental Quality 2010; Oklahoma Water Resources Board 2011).

Though conflict still surrounds groundwater management in Oklahoma, the state's rights-based institutions allow for conflict resolution through voluntary market transactions. Agricultural producers with permitted water rights can sell or lease groundwater to other beneficial uses, including municipal, industrial, and environmental. Compared to the protracted litigation that characterizes groundwater disputes in basins where the rights are poorly defined, Oklahoma's clear and complete allocation system allows for both administrative and resource efficiency.

Conclusion

Approaches to groundwater allocation have traditionally begun with central management because policy analysts have assumed that the definition and enforcement of rights are not feasible. As the value of water rises, however, additional efforts will be devoted to capturing the value of groundwater. Will these efforts be devoted to rent seeking from bureaucratic managers or to defining and enforcing rights that will encourage efficient, decentralized management? The evidence from Oklahoma and the Tehachapi Basin suggests that privatization has improved allocation at reasonable costs. Yet bureaucratic restrictions on pumping from stocks and on transfers continue to thwart further efficiency gains.

The basic proposal put forth by Smith (1977) for establishing both stock and flow rights remains a model for groundwater privatization. Where geohydrology results in third-party impairment, some restrictions on transfers might be necessary. Experiments with unitization of groundwater basins could go beyond the Smith approach to internalize all pumping interconnections and optimize extraction. The fact that environmental groups are proposing a property rights approach to groundwater management suggests that the time is right for further institutional reform that will facilitate groundwater markets.

10

YOU CAN'T KEEP A GOOD MARKET DOWN

Reports of water shortages, conflict, and crisis come from all over the world. In October 2010, water levels in Lake Mead in Nevada dropped to their lowest level since it was first filled in 1936. If this trend continues, lake levels would drop below the level of the intake pumps that supply Las Vegas, cutting off 90 percent of the city's water supply (Barringer 2010). In California, in 2009, thousands of farmers and farm workers marched in protest of the federal government's decision to cut water deliveries in order to supply flows to the endangered delta smelt. In North Carolina, water users organized a grassroots campaign to fight a proposed rate increase needed to recover the costs of infrastructure and efficiency improvements (Murawski 2011). In Mumbai, India, because of severe drought, municipal water officials cut supplies to residents. Massive protests followed, ending with the death of one citizen and numerous injuries to others as police beat back protesters to restore order. And in Australia's largest river system, the Murray–Darling Basin, 20 of the 23 catchments are considered to be in "poor" to "very poor" ecosystem health, threatening the capacity of rivers, streams, and wetlands to support native flora and fauna (Davies et al. 2008).

Politics Instead of Markets

This sampling of stories illustrates the importance of changing the institutions that we use to manage water on the "Blue Planet." The fundamental problem is that quantity demanded exceeds quantity supplied because markets are not allowed to equilibrate the two, which in turn leads to conflict and political management. On the supply side, political solutions require increases in storage and delivery, both of which are financially and environmentally costly. On the demand side, political solutions require controlling consumption through water use restrictions, mandates for low-flow fixtures, rationing, and subsidy programs aimed at improving delivery and use efficiencies. Although water-use regulations and conservation programs may be effective in the short term, they do little to correct the fundamental cause of over-consumption, namely prices that do not reflect scarcity values.

Las Vegas

The water crisis in Las Vegas demonstrates many of the institutional shortcomings as well as the inability of governmental regulations to square demand and supply. Las Vegas' primary water supply is Lake Mead, behind Hoover Dam. Drought has reduced lake levels so much that the state of Nevada fears reservoir levels will fall below present intake valves. To address the problem, construction began on a new, lower intake valve in 2008 that will allow the state to pump water even if lake levels continue to drop. The estimated cost of completing the new intake valve is $778 million (Southern Nevada Water Authority 2011), with no guarantee that the same situation will not unfold again in the future. In addition to the new intake valve, the state water authority has proposed a multi-billion dollar project that would pump groundwater in the northeastern part of the state and pipe it more than 300 miles to the Las Vegas Valley (Southern Nevada Water Authority 2010).

In addition to supply augmentation, the city has also adopted demand-side controls such as water-use restrictions and conservation programs. Las Vegas homeowners are no longer permitted to plant new grass in their front yards and those with existing lawns face strict watering schedules, especially in summer when all watering is prohibited at times. Other restrictions include banning restaurants from serving tap water unless requested by the customer, restricting the washing of cars both at home and at commercial car washes, and limiting water use for swimming pools and hot tubs (*U.S. Water News* 2006). To enforce its regulations, the Las Vegas Valley Water District maintains a separate enforcement staff to police water uses that are "prohibited and declared unlawful."[1] In 2009, the water police investigated over 6,200 cases of wasteful water uses and issued nearly $100,000 in fines (Las Vegas Valley Water District 2011). In addition to water-use restrictions, the Southern Nevada Water Authority has spent $147 million to encourage landowners to "xeriscape"[2] their lawns (Barringer 2010).

To be sure, the city of Las Vegas understands the importance of prices and has reluctantly tried to make prices more accurately reflect the scarcity of water by implementing an increasing block pricing schedule. This means that higher rates are charged for greater amounts of additional water use. However, the city's rates are comparatively low, especially given the fact that Las Vegas is in a desert. A family of four in Las Vegas using 100 gallons per person per day pays less than a $1 per day, compared to the same family in Atlanta, Georgia, where precipitation is 13 times greater and who pays twice as much for the same volume. Only Jacksonville, Florida, has lower rates (Walton 2010) and it is far from a desert. Given its low water rates, Las Vegas has the highest per capita water consumption among cities with increasing block rate schedules (Walton 2010).

Getting government officials to increase water rates to better reflect scarcity, however, is particularly challenging. Raising rates spur protests and can be political suicide for county officials. For example, in Merced County, California, in 2010, the mayor and one councilwoman were voted out of office in a recall election because they supported increased water and sewer rates (North 2010). The aversion to higher water prices forces water utilities to operate on revenues insufficient to cover long-term infrastructure costs. In the United States, revenues from water rates are insufficient to cover the cost of service in 61 percent of drinking water utilities and 67 percent of wastewater utilities (U.S. GAO 2002). As a result, there is an estimated funding shortfall of $224 billion in the United States needed to maintain infrastructure (U.S. EPA 2002).

Bolivia

In other countries, rate increases have literally been deadly. In early 2000, in Cochabamba, Bolivia's third largest city, thousands of citizens stormed the streets in protest of the privatization of water services and rate increases. The government sent in over 1,000 police, armed with tear gas and live ammunition to retake control of the city. Before ending in April, one protester was shot and killed, more than 175 were injured, and property damages topped $20 million (Schultz 2003).

The riots broke out in response to subsequent water rate increases following the city's concession with Aquas del Tunari (AdT), a private water services firm. The main impetus for selling to the private sector was the poor performance of the municipal utility company, SEMPRA, which had accumulated roughly $30 million in debt and was one of the worst performing public water utilities in Latin America (Nickson and Vargas 2002). For decades SEMPRA charged customers water rates insufficient to cover costs, much less expand service to the roughly 40 percent of the population that lacked access to public water supply and the 50 percent of the population without sewerage service (Bonnardeaux 2009).

The rate increases following privatization were guaranteed by the government's contract and used to cover the cost of securing new water sources, expanding service, paying down past debt, and a 16 percent rate of return on capital as agreed to by city officials (Nickson and Vargas 2002). On average, water rates increased 35 percent but the increase varied widely among customers, from less than 10 percent for small and low-income customers to more than 200 percent for industrial and commercial users (Bonnardeaux 2009). In the first two months, AdT improved supplies by roughly 30 percent, in part by reducing leakage rates that were estimated to be as high as 50 percent, and in part by improving technical efficiencies (Bechtel Corporation 2005). Customers, particularly low-income households who had been accustomed to frequent disruptions in service and rationing, were provided with better access to water and consumed more, thus saw their water bills increase substantially (Israel 2007).

The riots resulting from rate increases induced the government to declare a state of emergency and to cancel its concession contract with AdT. With control of the water utility back in state hands, half of the residents served by the utility do not have service while many of those with service only have running water a few hours a week (Shultz 2010). Those households without piped water still rely on private water vendors and pay prices higher than those with municipal service (Israel 2007). Despite the revolts and pressure to improve public service, the "company that manages it remains riddled with corruption, mismanagement, and inefficiency—a source of graft for the city's mayor and the union that represents the company's workers" (Schultz 2010).

Fueling political pressures to keep water rates low is the argument that water is a "human right" (Barlow 2007, 130). As such, it should be free to all. Though declaring water a basic human right has a nice moral tone, it begs two questions: how much water do people have a right to and who will pay to supply the water to which there is a human right? Pricing water at zero or even low levels ensures that the quantity demanded will remain high and continue to increase, if only with population growth, and that there will be insufficient revenues to fund infrastructure and operating costs.

Mixing Politics and Markets

After a century of massive water projects subsidizing water use, the U. S. Department of Interior's Bureau of Reclamation is one surprising example where water marketing is being tried. In 1995, the bureau outlined a process for transferring ownership in federal projects to non-federal entities. Title transfers put ownership of the resource in the hands of those that directly benefit from the project such as contractors, irrigation districts, and water users. Doing so reduces operational and management costs and improves efficiencies. In the words of the Bureau, it is "a recognition of Reclamation's commitment to a Federal Government that works better and costs less" (U.S. Bureau of Reclamation 2006, 35). By 2006, 19 project facilities had been transferred and five others had been approved. The process is slow and can be costly, but efforts are underway to streamline transfers and reduce costs. Legislation introduced in 2008, but not yet approved, would establish a program to identify additional projects for transfer and authorize the interior secretary to transfer facilities without congressional approval, thus avoiding a time-consuming and expensive process (Johnson 2008).

The bureau also understands the importance and capacity of markets to reallocate water within public projects. Introduced in 2009, the Water Transfers Facilitation Act, if passed, would expand water trading opportunities in the California's Central Valley Project. In support of the bill, Michael L. Connor (2009), commissioner of the Bureau of Reclamation, stated:

> When they are done right, water transfers move water from willing sellers to willing buyers in transactions that can improve economic well-being, increase efficiency in water use, and protect against negative externalities. There are many situations where water transfers during periods of drought

can be used to ensure that available water is used in areas where it is most needed, and [the Water Transfers Facilitation Act of 2009] is aimed at facilitating these efficient water transfers. We recognize the potential of voluntary water transfer as a mechanism to increase flexibility into our water management system and respond to changes in available water resources.

The concept of water marketing has come a long way since economists and policy analysts first began advocating the idea. Trades between agricultural users and cities are more commonplace and involve ever-increasing quantities of water. As we have seen, environmentalists are leasing and purchasing water for instream uses such as fish and wildlife habitat or recreation in many parts of the American West. The U.S. Environmental Protection Agency, together with farmers, municipalities, and conservationists have developed trading programs that find the cheapest way to improve water quality, as in North Carolina's Tar-Pamlico Sound. Eastern riparian states are moving in the direction of clarifying and quantifying water-use rights that could lead to more water markets in that region. And, instead of protracted court battles that might eventually net more water for a tribe but leave them with no capital to develop the water, Indian tribes are negotiating with other water users and the states. The Shoshone-Bannock Tribe in Idaho, for example, now operates a local water bank that facilitates the leasing of water among users, reallocating water to higher valued uses, including instream flows.[3]

Going Global

Water marketing is not just confined to the United States. Indeed other countries are facing the same or worse shortages. Nations around the world—prosperous and developing, water rich and water poor—are facing the challenge of balancing water demands with water supplies using markets. One of the reasons water markets are on the rise is that purely political solutions do not have a very good track record. Especially when problems arise in the developing world, it is not an option to simply plow more money into increasing supply or ration use through restrictions on access to potable water for drinking and domestic uses. As a result, water shortages are becoming the mother of institutional invention. As the following two international examples illustrate, creating well-defined, secure, and transferable rights is not easy, but it is absolutely necessary if water markets are to help equilibrate demand and supply.

Water Markets Down Under

Australia, like a growing number of other nations facing increasingly scarce water resources, has adopted a market-based system to allocate water. In the states of New South Wales and South Australia, severe drought during the 1983–84 growing season meant farmers received 10–20 percent of their annual water entitlements, an amount that, for many, was below what was needed for planting crops (Haisman 2005). Facing a collapse of agricultural production, the South Australian state government initiated a temporary trading program that allowed farmers to aggregate their water allocations through trade, expanding access to water for farmers while providing others an opportunity to sell water that would otherwise have been unused. Since then, other Australian states, territories and the Commonwealth as a whole, have undertaken considerable steps to expand markets and improve water-use and allocation efficiencies. Today, trading of temporary allocations and permanent entitlements is possible throughout most of Australia.

The path to today's Australian water markets has been long and hard. Unlike the western United States where a system of tradable water rights was adopted early on, governance of water resources in Australia has, historically, more closely followed that of the riparian doctrine. Beginning with the early settlers in the nineteenth century and through much of the twentieth century, water rights were tied to landownership and were not well defined, divisible, nor transferable. Moreover, there was uncertainty about access to water over time, which made entitlements less secure and diminished incentives for investment in new water developments and use technologies (ACIL Tasman 2004).

Limitations on trade were of little concern as long as water was abundant enough to meet all demands. But after more than a century of water development, erratic precipitation, and growing demands for offstream as well as instream uses, many rivers are now over-allocated and water shortages and conflict are not uncommon (Quiggin 2001). This brought about a realization that the institutions governing water were no longer consistent with growing demands in both type of use and quantities.

As water resources became increasingly scarce, opportunities arose for economic gains from trade. Absent the legal provisions that would have provided for formal trading of water entitlements, farmers captured these gains indirectly through informal markets. Entitlements were transferred through duality of ownership, or licence stacking whereby farmers purchased two landholdings and transferred the water entitlement from one to the other. Despite the added costs and idle land, their willingness to undertake the increased transaction costs associated with such transfers suggests the gains that were available through water trades.

Water reforms followed, beginning in the 1980s with the first provisions for trade in New South Wales and South Australia. Broader reforms came in 1994, when the Council of Australian Governments (COAG) endorsed the Strategic Framework on Water Reform that laid the foundation for a nationwide transition to formal water markets. Central to the reforms were provisions for defining water entitlements, severing water claims from land, and incorporating environmental flows into water management plans. State and territorial governments were tasked with redefining entitlements in terms of ownership, volume, reliability, transferability, and, if appropriate, quality, while also developing plans to manage water for instream applications (Bennett 2005, 78). To halt further declines in stream flows, the Murray–Darling Basin Ministerial Council implemented a basin-wide cap on withdrawals in 1995, putting a stop to new diversions (MDBC 2008).

The reforms in the 1990s resulted in "considerable progress toward more efficient and sustainable water management" (COAG 2004). However, improvements to trade efficiencies and provision of instream flows were still needed (Brennan and Scoccimarro 2000; Garry 2007). To address these issues and to expand and reinforce earlier reforms, state, territorial, and national governments signed the National Water Initiative (NWI) in 2004. The goals of the NWI include improvements to water management for environmental purposes, overcoming barriers to trading water entitlements, and improving the security and clarity of water entitlements (COAG 2004). Through these reforms, the NWI is expected to improve the effectiveness and efficiency of water markets by improving access to information on trades and entitlements, streamlining trades within and between states or territories, providing protection from injury to third parties, and increasing incentives for investment by improving the security of entitlements (Garry 2007). According to the National Water Commission (2010), progress in meeting these goals has so far been good, but the problems of over-allocated streams and insufficient environmental flows still remain.

In the case of Australia, the actions of successive governments have complicated trading and the provision of instream flows (Murray–Darling Basin Authority 2010b). Decades of over-allocating water without a clear accounting of supplies has left streams dry and left many entitlement holders with uncertain access to water, especially in dry years.

To overcome these challenges, Australian governments are employing a mix of markets and investment in infrastructure. In 2008, the national government announced its Water for the Future initiative, a 10-year program aimed at improving water-use efficiencies and environmental flows, as well as providing risk abatement for climate change. The program will invest $4.4 billion in infrastructure and new technologies to improve water-use efficiencies, helping farmers maintain or improve productivity with less water. An additional $3.1 billion will be used to acquire water entitlements from willing sellers. The acquisitions will reduce overall diversions and improve the reliability of remaining entitlements. In addition, the program provides $56 million in funding to develop a National Water Market System. The initiative will "strengthen Australia's water market through efficient management of improved state and territory water registers, water transactions and the availability of market information"—lowering transactions costs and improving allocation efficiencies (Australian Department of Sustainability, Environment, Water, Population and Communities 2010).

Since the first temporary trades in the 1980s, market activity has increased considerably, especially in the Murray–Darling Basin (MDB), the largest and one of the most important ecological as well as economic river systems in Australia (Murray–Darling Basin Authority 2010a). In the late 1980s and early 1990s, trading volumes averaged roughly 146,000 acre-feet and 468 trades per year (Sturgess and Wright 1993, 23–24), compared to over 3.3 million acre-feet and more than 25,000 trades in 2009 and 2010 (National Water Commission 2010, 28–31). Trading activity in the MDB now accounts for roughly 93 percent by volume of all water traded in Australia and it is expected to continue to increase (National Water Commission 2010, 28).

Tradable water rights have provided farmers with a means of adapting to uncertain supplies, while creating incentives for improved water-use efficiencies. As competition for water has increased, so too has its value. In a market setting like the MDB, scarcity values are reflected in the price of entitlements and allocations, and higher prices have created incentives for improved water-use efficiencies and investment in water-saving technologies. As a result, farm productivity has been increasing by roughly 3 percent annually (Australian Bureau of Statistics 2007, 20), despite reduced inflows into the basin (Murray–Darling Basin Authority 2010a). Grafton et al. (2010, 18) note the importance of trading in the MDB in the face of increasingly uncertain water supplies.

> The principal beneficiaries of water trading in the MDB have been perennial farmers who irrigate orchards and vineyards and who, despite having high reliability water entitlements, have found that their assigned seasonal allocations were less than they expected and needed. Without the ability to purchase seasonal allocation (temporary) water over the past four years, many of their vineyards and orchards would have suffered major harm or died in the present drought. Sellers of seasonal water have also benefited as the increased volume of sales, at high water prices, provide an important source of income . . . Higher prices have encouraged investments in on-farm water efficiency. The ability to trade and to adjust the volume and mix of high and low reliability water entitlements to reduce risks of insufficient water supplies has also permitted investments in perennial agriculture that may otherwise not have been contemplated.

Australia's transition to markets illustrates not only the benefits to and capacity of trade to more efficiently and effectively allocate scarce water, but also the challenges associated with adopting a market-based system. Enacting legislation that severs water from the land and that provides for trading among various uses and users is a necessary condition, but not sufficient to ensure a functioning market. The fundamental components of a market such as well-defined, secure, and divestible entitlements must also be present.

The Chilean Experiment

Chile is a popular case study among proponents of water markets, but its water market record is mixed. As geographer and legal scholar Carl Bauer (1997, 639) explains, the "real lesson of the Chilean experience is that establishing markets in water resources is harder than it may seem . . . markets are not simple, automatic, or self-maintaining mechanisms: how they operate depends on wider legal and institutional frameworks, political and economic conditions, and geographic context."

The explicit endorsement of water markets throughout the country's water code made Chile the international model for treating water not only as private property, but as a fully marketable commodity (Bauer 2004). Despite this reputation, active trading has been documented in only a handful of river basins (Hearne and Donoso 2005). The absence of trading elsewhere has led some commentators to question whether the success of Chile's water market is overstated (Bauer 1997) or altogether illusory (Barrionuevo 2009).

Chile's water law has not always embraced free market principles. Before the 1980s, the central government played a significant and sometimes technocratic role in water allocation. Under the country's 1967 Water Code, the government expropriated vast amounts of water rights, prohibited private water transactions, and created the Directorate General of Water, a state agency with expansive authority to reallocate water based on technical standards of "rational and beneficial use" (Bauer 1997, 642). This centralized approach to water allocation coincided with widespread expropriation and redistribution of private agricultural land that occurred during the agrarian reform of the 1960s.

Not until the 1973 military coup did the country's economic paradigm shift from "one where the state must ensure the optimum allocation of resources, to one where market forces dictated efficient allocation" (Hearne and Donoso 2005, 57). But even after much of the expropriated land had been returned to private ownership, the government continued to treat water rights as revocable administrative concessions. As Bauer (1997, 642) noted, the "legal insecurity of water rights discouraged private investment in water development or management, and the system's inflexibility prevented transfers to higher-valued uses. Water rights titles were especially uncertain because after 1967 neither they nor their transactions were recorded in official property registries."

To facilitate water conservation, investment, and trading, Chile needed institutional change reflecting the country's new economic paradigm. That change occurred with Chile's 1980 constitutional reforms and 1981 National Water Code.[4] Rather than mandating the creation of a water market, these legal reforms were an attempt to create the institutions necessary for spontaneous water trading (Bauer 2004). Specifically, the 1980 constitutional reforms recognized water rights as private property separate from landownership and secure from uncompensated government expropriation (Gazmari and Rosengrant 1996).

The 1981 Code created permanent water-use rights, separated them from landownership, and allowed the use rights to be bought, sold, leased, and mortgaged in the same manner as real property (Hearne and Donoso 2005, 57). Together, these laws "greatly strengthened private property rights in water, increased private autonomy in water use, and favored free markets in water rights to a degree that was unprecedented both in Chile and in other countries" (Bauer 2005, 150). In addition to creating secure and transferable water rights, the 1981 Code also significantly limited the authority of the General Water Directorate (Direccíon General de Aguas or DGA). Specifically, the Code required the DGA approve all applications for new water rights if the water is physically and legally available.[5] It also exempted water rights holders from any use requirements, use specifications, or change of use approvals (Bauer 2004). In stark contrast to the beneficial use and use-it-or-lose-it principles of the prior appropriation doctrine, Chile's 1981 Water Code truly embraced private ordering as the exclusive means of determining water's highest and best use.

Even with secure and transferable water rights, active trading only occurs in a handful of Chile's river basins (Donoso 2006). Two factors explain this: (1) the economic conditions that necessitate active trading do not exist throughout the entire country (Bauer 1997); and (2) though the 1981 Water Code created secure and transferable water rights, it failed to define those rights with sufficient clarity to minimize transaction costs. Neither of these factors supports the claim that Chile's water market has failed, but rather that Chile's water market has unrealized potential.

The first factor limiting the volume of trades in Chile's water market is scarcity, or a lack thereof. In sparsely populated southern Chile, water is abundant and water demands are few (Hearne and Donoso 2005). Without competing demands for scarce water resources, there is no impetus for trading or, for that matter, defining private property rights in water. Water trades are similarly rare in northern Chile, for quite a different reason. There, water is extremely scarce. The Atacama Desert is one of the driest places on Earth, with some locations having never recorded a precipitation event (Vesilind 2003). What little water does exist has already been put to use in high-value mining applications (Bauer 1997) and there is no appreciable demand from other sectors to support an active market. However, intra-sector demand combined with ever-present drought conditions has prompted the region's mining operations to increase water-use efficiency threefold over the last 20 years (Williams and Carriger 2005).

Only in Chile's central valley system do competing demands for scarce water resources support frequent trading. The Limarí River basin in north-central Chile has a particularly active market due in large part to increasing irrigation demands, well-organized canal associations, and three large storage reservoirs (Hearne and Donoso 2005). These reservoirs allow banking and transfer of specific volumes of water, which helps reduce transaction costs (Bauer 1997). Central Chile, which contains the country's major industrial and urban areas, also boasts an active water market. There, high-value export crops such as fruit, vineyards, and vegetables compete with industrial and municipal water demands (Hearne and Donoso 2005). Hydroelectric operations have also placed demands on Central Chile's water resources, increased their market value, and driven trades.

As Donoso (2006, 162) explains, "markets are more active in those areas where the water resource is scarce with a high economic value." Where water supplies greatly exceed water demands as is the case in southern Chile, and where water has already traded to its highest-valued uses as it has in northern Chile, water trading is less common. The thinness of the market in these regions is not regrettable, nor does it suggest Chile's water market has failed. Rather, it simply suggests that the current economic and hydrologic conditions do not support frequent trades. Only when these conditions change will more trading occur.

The second factor limiting the volume of trades in Chile's water market is uncertainty of rights created by the 1981 Water Code's recognition but not recordation of pre-existing use rights. Specifically, the 1981 Water Code declared valid all use rights recognized or created by previous laws regardless of how they had been acquired[6] or whether they had been formally registered (Figueroa del Río 1995). The purpose of this provision was to eliminate the uncertainty over existing titles by declaring a presumption of ownership in favor of current users (Bauer 1997). However, without a corresponding requirement of recordation, no certainty of ownership was achieved.

With the portion of unregistered rights estimated to be between 60 and 90 percent (Dourojeanni and Jouravlev 1999), buyers and sellers face enormous uncertainty over the validity of use rights. As Bauer (1997, 647) explains, with "an unknown number of legally valid rights which in theory could be asserted at any time . . . the principal legal obstacle to trading is the continuing uncertainty of many titles." It is unclear whether the 1981 Water Code's requirement that all rights be officially recorded before they could be transferred created an incentive for recordation or a disincentive for trade. In either event, the ever-present possibility of overlapping claims and third-party impacts generates significant transaction costs and thus limits the efficiency gains of market allocation (Donoso 2006).

Unlike the economic and hydrologic conditions which limit water trades to central Chile, the institutional issue of poorly defined water rights can and should be remedied through legal reform. Mandatory recordation of all water rights in a central registry is ultimately essential and particularly important for the traditional rights that predate the 1981 Water Code. Requiring recordation of rights as a prerequisite to trade is an elegant way to parse legitimate and illegitimate claims, but the Chilean government should establish a recordation deadline in order to eliminate the uncertainty created by vested rights unknown to transacting parties. Applying the 1981 Water Code's taxonomy for newly established water rights[7] to pre-1981 claims is a low cost but important means of standardizing recorded rights. Adjudication of conflicting rights is also necessary. Though costly, these efforts to more clearly define Chilean water rights are essential to the market allocation process and rational as Chile's water resources become scarcer.

Though Chile's 1980 constitutional reforms and 1981 Water Code established private, secure, and transferable water rights, the country's water market has not met expectations (Bauer 1997; Donoso 2006). The variable performance of Chile's water market is not surprising, nor does it evidence a failure of water markets. Rather, it suggests that trading water only makes sense in the presence of competing demands and clearly defined rights.

Getting from Here to There

According to the United Nations, by 2025, 1.8 billion people will be living in countries or regions with absolute water scarcity and two-thirds of the world population could be under conditions of water stress (United Nations Environmental Programme 2007). By 2050, as many as 4.8 billion people could face severe water scarcity and the impact on global economies could be considerable. The International Food Policy Research Institute estimates that by 2050, as much as $63 trillion in global GDP could be at risk because of water stress (Veolia Water 2011).

Going forward, there is no doubt that water scarcity will increase the frequency and severity of water shortages, conflict, and crises, at least in the short term. Understandably, with global populations expected to top 9 billion by 2044, increased urbanization that will concentrate water demands in regions with relatively fixed supplies, and expectations of higher per capita consumption, water demands could increase considerably. On the supply side, climate change and growing demands for instream flows and improved water quality will compete with water for consumptive uses, potentially widening the disconnect between water supplies and demands.

To be clear, predictions of water shortages and crises are predicated on business as usual scenarios; that is, what will happen if we do not change the way we manage our water resources. Although unlikely to come to fruition, such dire predictions are still useful. For one, they communicate to the international community a sense of urgency and the importance of developing long-term and self-sustaining management strategies. Second, and perhaps more importantly, they are indicative of the problems or shortcomings in conducting business as usual, which is demonstrably incapable of adapting to current shifts in demand and/or supply, let alone uncertain future changes.

When the gains from trade become large enough, however, it is hard to keep a good market down. When water from a federal project such as the Central Utah Project is delivered to farmers at $30 per acre-foot to produce crops, but at a cost to the taxpayer of $161 per acre-foot (U.S. Department of the Interior 2009), it is obvious something is wrong. When Santa Barbara, California, is building a desalination plant to produce potable water at a cost of $1,600 per acre-foot while farmers are using water to irrigate crops where it is worth less than $100 per acre-foot, the potential for mutually beneficial trades cannot be ignored. With an order of magnitude difference in value, both sides of a water market transaction can gain substantially. When environmental groups realize that the transaction costs associated with using the regulatory process are so high that amenity values are lost while lobbyists and politicians play games, water markets in which environmental consumers pay other users to reduce consumption become expedient. Transaction costs are much lower when willing buyers and willing sellers are seeking common ground for mutually beneficial trades. Each such trade moves us farther along a water market path from which it will be difficult to reverse.

Indeed, as this book has described, markets and the necessary institutions to support trading have expanded considerably in recent decades. Courts and legislatures in the United States and other nations have amended water laws to sever water rights from landownership and remove legal barriers to trading those rights. Combined with expanded beneficial use standards and appropriation laws that legally recognize instream flows, water markets are now helping to enhance stream flows for recreation and improve habitats for fish and wildlife. Where these legal changes have occurred, local, state, and in some cases, national governments, are working to better define and enforce water rights and entitlements, reducing information costs and third-party effects from trade, while improving water accounting and inventory data. And where it is now possible to capture gains from trade, buyers, sellers, and entrepreneurs are finding ways to reduce transaction costs and streamline transfers. Private conservancies are acquiring water for environmental uses and working with state agencies to improve the administrative process and trading efficiencies. And private water brokers and research companies such as WestWater Research, LLC and Lotic, LLC in the United States and Waterfind in Australia are helping to connect willing buyers and sellers and reduce the transaction costs of trading.

The daunting challenge, of course, is to move away from the status quo where prices do not reflect scarcity and where politics dominate allocation. The United States has experienced nearly a century of federal dominance in water policy centered around massive projects to control flooding and navigation in the East and to "make the desert bloom like a rose" in the West. Roughly 84 percent of households receive their water from publicly owned utilities and 75 percent are served by public wastewater treatment plants (U.S. Department of Homeland Security and EPA 2007). First-world nations from Australia to Norway depend on governments for storage and delivery systems rationalized on the basis of economic development, with too little consideration given to fiscal or environmental impacts. Private individuals who have captured the benefits of water subsidies will not simply push themselves away from the trough, and politicians and bureaucrats who have enjoyed the power that accompanies command-and-control regulation will not readily relinquish that power. Moreover, proponents of governmental support for dams and ditches undoubtedly will find support among the citizens of less developed countries who feel they deserve the same largess that has been enjoyed by their rich neighbors. Therefore, when the World Bank or some other development agency proposes to dam the Zambezi River or to subsidize irrigation projects in the Middle East, fiscal and environmental arguments are likely to be swept aside. In short, water will gush uphill to political power.

The evidence presented in this book suggests that there is tremendous potential from tapping water markets. Some would say that water cannot be entrusted to markets because it is a necessity of life. To the contrary, because it is a necessity of life and so precious, it must be entrusted to the discipline of markets that encourage conservation and innovation. Unless distortions created by governmental intervention are corrected, however, water shortages will become more acute, crises will be inevitable, and the poorest people in the world will likely be the ones who bear the brunt of bad policies.

When this happens, it will be difficult to suppress market forces. It would be better if we could get legal impediments out of the way before crises worsen. By doing so, we can clear the way for entrepreneurs who will find ways to promote gains from trade through water-use efficiency. Water markets will not solve every water allocation problem, but just as Adam Smith's invisible hand of the marketplace has promoted "the wealth of nations," tapping water markets can enhance the water wealth of our planet.

NOTES

1   Why the Crisis?

1   "Unimproved drinking water sources" include unprotected dug wells, unprotected springs, carts with small tank/drum, tanker truck, surface water (river, dam, lake, pond, stream, canal, irrigation channels), and bottled water. "Unimproved sanitation facilities" refers to sanitation practices that fail to ensure hygienic separation of human excreta from human contact.

2   Lomborg suggests the increase in per capita water consumption might be more a reason for celebration than concern. Specifically, the increase in per capita water use reflects a 50 percent increase in water use for agriculture, allowing irrigated farms to feed us better and to decrease the number of starving people, while in developing countries, better access to clean drinking water and sanitation reduces incidence of water-borne diseases.

3   According to Gleick (1996), approximately 97 percent of the world's water is saline. Of the remaining 3 percent that is fresh water, 68 percent is locked up in glaciers and ice and 30 percent is groundwater (quoted at USGS 2009, accessed November 17, 2009).

4   "Water stress" is any amount of usable water below 1,700 m³ per capita per year.

5   An acre-foot of water is equivalent to 325,851 U.S. gallons (1,233 m³)—a volume sufficient to cover one acre of land to a depth of one foot.

6   Eutrophication is the over-enrichment of water by nutrients such as nitrogen and phosphorus. Hypoxia (or oxygen depletion) is a symptom of eutrophication.

7   This includes major water bodies such as the Chesapeake Bay, the Baltic Sea, the Gulf of Mexico, and the Tampa Bay.

8   If Waukesha County had lain wholly outside the basin, it would have no prospect of diverting water from Lake Michigan as such diversions were the primary target of the compact.

9   For more discussion of the prior appropriate system, see Chapters 4 and 5.

2   Cheaper Than Dirt

1   For examples, see Bailey (1995); Simon (1995); Lomborg (2001).

2   Here we focus on costs not accounted for in market transactions and recognize that it is also possible for benefits to be ignored if some parties reap value from market

transactions for which they do not pay. In the benefit case, private benefits will be less than social benefits and production of the good or service will be less than optimal.

3 For an excellent introduction to public choice, see Mitchell and Simmons (1994).
4 For one of the first discussions of rent seeking, see Tullock (1967).

3 Who Owns the Water?

1 *Red River Roller Mills v. Wright*, 30 Minn. 29, 15 NW 167, 169 (1883).
2 In some of the states that have established a permit system to allocate water among riparian users, forfeiture from non-use is possible. Three years is a common forfeiture period (Caponera 2007, 127).
3 The doctrine of prior appropriation is that body of state law under which individuals can secure a water right by diverting water from a watercourse and applying it to a beneficial use. One's prior right to water is determined by the date on which he or she first applied the water to a beneficial use, known as the priority date. In times of shortage, older, or senior, rights must be satisfied before newer, or junior, appropriators may take their water. The doctrine places no limitation on the place of use, and appropriative rights are transferable. Nonuse of a water right may be considered abandonment and result in its forfeiture. Although the doctrine arose in state constitutional and case law, eventually it was codified in state statutes and tailored to suit the needs of each individual western state.
4 *Hoffman v. Stone*, 7 Cal. 46, 48 (1957).
5 *Butte Canal and Ditch Co. v. Vaughan*, 11 Cal. 143, 152 (1858).
6 *Weaver v. Eureka Lake Co.*, 15 Cal. 172, 175 (1860).
7 For a discussion of the this problem, see Anderson and Hill (1990).
8 The climate of landownership in California was conducive to development corporations because speculators had acquired thousands of acres under the federal land laws such as the Desert Land Act of 1877. Although the federal legislation was designed to convey land to individuals in 160-acre parcels, speculators manipulated the law through the use of dummy entrymen and bribery to appropriate considerably larger tracts. As land values increased, these individuals sold and developed their holdings in various ways, including the use of development corporations (Dunbar 1983, 31–32; Worster 1985, 98, 100–2).
9 The Tempe Irrigating Canal Company, for example, organized in 1870, constructed irrigation works to distribute 11,000 miner's inches of water from the Salt River in Arizona. The Big Ditch Company was organized to purchase a project constructed by the Minnesota and Montana Land and Improvement Company west of Billings, Montana in 1900 (see Dunbar 1983, 29–30).
10 Utah passed irrigation district legislation in 1865, but it did not give districts the power to issue bonds, which left them powerless to raise revenue for construction of new projects (Huffman 1953, 75; Worster 1985, 79).
11 The basis for including in the district all land located within its geographical boundaries was that all property would increase in value once the surrounding lands were irrigated. Some states provided for exemption of property from the district if its owner applied to withdraw it (Wyoming), the land would not be benefitted by the district (California), or was located in a town or city (Oregon) (Chandler 1913, 135).
12 Mass. Gen. Laws ch. 216, §4 (2002).
13 Va. Code §62.1–44.15:22(B).
14 Water that is not returned to the immediate water environment is considered consumptive. Generally this would include water lost through evapotranspiration, water incorporated into products (e.g., crops), or consumed by humans or livestock.

4 Water Is for Fightin'

1 Howitt et al. (2009) estimated lost revenues of $1.2–1.6 billion to the Central Valley agricultural industry from a 90 percent reduction in State Water Project deliveries and

a 100 percent reduction in Central Valley Project deliveries—the predicted impact of drought and legal restrictions for the 2009 growing year. The authors assumed irrigators could mitigate much of the short-term economic impact through stress irrigation, increased groundwater pumping, and land use changes. Without such short-term mitigation and with a complete loss of Delta deliveries, the economic value estimates would presumably increase.

2   Dunbar (1983) recounts numerous examples of irrigators resisting attempts of canal companies to charge perpetual right fees in addition to the annual operation and maintenance charges. Though some of this resistance took the form of legislation prohibiting perpetual right fees, many of the water supply contracts were simply renegotiated.

3   For an example of the infrastructure deterioration possible under public management, see Minardi (2010).

4   *Jennison v. Kirk*, 98 U.S. 453, 461 (1878).

5   Chapter 5 provides a thorough discussion of the prior appropriation doctrine.

6   3 Cal. 249, 252 (1853).

7   *National Audubon Society v. Superior Court of Alpine County* (Mono Law case), 33 Cal. 3d 419, 658 P. 2d 709 (1983).

8   For a complete discussion of the history and implications of this case, see Libecap (2006).

9   In the 15 years preceding the injunction, the city of Los Angeles diverted an average of 83,000 acre-feet per year. In October, 2009, the lake level elevation was 6381.7 feet, roughly 10 feet below the target level. Arnold (2004) estimates target lake level targets will be met by 2021.

10   As stated in *United States v. Appalachian Elec. Power Co.*, 311 U.S. 377, 426–27,

> [I]t cannot properly be said that the constitutional power of the United States over its waters is limited to control for navigation. . . . In truth the authority of the United States is the regulation of commerce on its waters. . . . The authority is broad as the needs of commerce. . . . The point is that navigable waters are subject to national planning and control in the broad regulation of commerce granted the Federal government.

11   For a more detailed discussion of the federal government's power over the waterways of the United States, see Washburn (1986).

12   The navigation servitude is generally considered a subset of Congress' broad power to protect interstate commerce under the Commerce Clause.

13   *The Daniel Ball*, 10 Wall. 557, 563, 19 L.Ed. 999 (1871).

14   For additional information on navigability, see Baughman (1992, 1028).

15   S. 787. Clean Water Restoration Act. 111th Congress, 2009–2010.

16   *Winters v. United States*, 207 U.S. 564 (1908). For an excellent discussion of this case and the impact it has had on prior appropriation claims, see Smith (1992).

17   *Winters v. United States*, 143 F. 740, 749 (1906).

18   The magnitude of the rent-seeking potential in the reclamation system was discussed in Chapter 4 in the context of extensions of the reclamation subsidy and circumvention of the acreage limitation.

19   Prior to issuing the principles, the various regions of BuRec had been inconsistent in dealing with transfers of project water, generating confusion and uncertainty about the legality of such transactions.

20   Third parties are defined as "those entities who may have some identifiable interest in the exchange, and would have a legal standing in an adjudication process in an appropriate State forum." U.S. Bureau of Reclamation (2010, 4).

21   The CVPIA was part of the larger Reclamation Projects Authorization and Adjustments Act, which addressed reclamation issues, generally, and voluntary transfers of reclamation water, specifically, throughout the West.

22   The CVPIA also established a three-tiered pricing system for federal agricultural water which requires irrigators to pay increased rates for CVP water (up to the full cost of

any water above 90 percent of the contract total). While the increased prices will encourage conservation and reduce the subsidy long enjoyed by irrigators, unrestricted transfers of project water would still be more efficient, because "[t]ransfers at market determined prices would allocate water to highest value uses and provide the same incentives for water conservation as does the three-tiered system." Gardner and Warner (1994, 7).

23   The CVPIA set aside 800,000 acre-feet of Central Valley Project water per year (about 13 percent of the total) for fish, wildlife and environmental restoration. By allocating such a substantial amount of project water to environmental purposes, the act elevates fish and wildlife to a status above the uses represented by water contracts, because the former are not required to compete in the market for water resources like municipal, industrial, and agricultural uses (Gardner and Warner 1994, 7). The act also established a $50 million restoration fund to carry out environmental and wildlife mandates included in the act. The money for the fund will come from surcharges on CVP water users (CQ Almanac 1992, 269).

24   On October 7, 2009, following years of sustained drought in California's Central Valley, Senators Dianne Feinstein and Barbara Boxer introduced a bill, the "Water Transfer Facilitation Act of 2009" (S. 1759), aimed at expediting Reclamation's review of water transfers by specified CVP water districts. In relevant part, the bill mandates certain "voluntary water transfers shall be considered to meet the conditions described in subparagraph (A) and (I) of section 3405(a)(1) of the Reclamation Projects Authorization and Adjustments Act of 1992." If approved by Congress, this legislation could alleviate the regulatory burden imposed by the Reclamation review, but only for those water transfers specified in the bill. More holistic streamlining of the review process will increase the transferability of Reclamation water and, consequently, improve the allocative efficiency of federal water.

25   For a thorough discussion of the impacts of federal water projects on agricultural efficiency, see chapter 12 of Gardner (1995).

5   Back to the Future

1   The Colorado Supreme Court articulated this rule in *Fuller v. Swan River Placer Mining Co.*, 12 Colo. 12, 19 P. 836 (1888), concerning the dumping of mining tailings and its impact on water users downstream, and again in *New Cache la Poudre Irr. Co. v. Water Supply and Storage*, 49 Colo. 1, 11 P. 610 (1910), concerning the change of a canal company's diversion point.

2   See *National Audubon Society v. Superior Court*, 33 Cal. 3rd 419 (1983).

3   See *Montana Coalition for Stream Access v. Curran*, 682 P.2d 163 (Mont. 1984); *Montana Coalition for Stream Access v. Hildreth*, 684 P.2d 1088 (Mont. 1984); *Bitterroot River Protection Ass'n. v. Bitterroot Conservation District*, 2008 MT 377.

4   *Union Mill and Mining Co. v. Dangberg*, 81 F. 73, 94 (C.C. Nev. 1897).

5   For a more complete discussion of why beneficial use made sense in the early West, see Anderson and Johnson (1986) and Sprankling (1994).

6   Montana Code Annotated §85-2-102(2) (1993).

7   Montana Code Annotated §85-2-104, passed in 1979 and repealed in 1985.

8   *Colorado River Water Conservation District v. Rocky Mountain Power Company*, 158 Colo. 331, 406 P.2d 798, 800 (1965).

9   Wash. Stat. 41-3-1001.

10   RCW 90.54.020(1); RCW 90.54.020(3)(a); RCW 90.22.010.

11   See Chapter 7 for a detailed discussion of instream flow laws.

12   Some states require consideration of whether the public interest will be impaired by the change or transfer (Colby et al. 1989, 718). Inclusion of these public interest factors or criteria in the transfer process is discussed later in this chapter.

13   Montana Code Annotated §85-2-402 (1993). Applications for changes in water rights must be approved by the state Department of Natural Resources and Conservation.

Additional criteria for changes to water rights include adequacy of the proposed diversion works, proof that no injury to the water quality of other appropriators will occur, and, when the proposal involves 4,000 acre-feet or more, application of public interest criteria.

14 The trend also embodies the standard reaction of many environmentalists to externalities, which is to internalize them through command and control regulation (see Huffman 1994).

15 CA Water Code §386.

16 ID Code §42-222.

17 Colo. Rev. Stat. §37-92-305 (4.5).

18 The Western Governor's Water Efficiency Working Group even went so far as to say that *benefits* to rural areas can result from water transfers away from rural areas: "If land removed from production as a result of a transfer was marginal and if the revenues from the transfer remain at least in part within the transferring area to modernize irrigation operations or to diversify the local economy, benefits can result" (Western Governors 1987, 109).

19 For example, the Metropolitan Water District of southern California acquired 106,100 acre-feet of salvaged water per year for the next 35 years by paying to line irrigation canals and funding other projects to improve water-use efficiency in the Imperial Irrigation District (IID). The irrigated acreage within the IID is unaffected by the agreement (see National Research Council 1992, 82–83, 243).

20 See Bates et al. (1993, 152–98).

21 See *National Audubon Society v. Superior Court of Alpine County* (Mono Lake case), 33 Cal. 3d 419, 189 Cal. Rptr. 346, 658 P.2d 709, *cert. denied*, 104 S.Ct. 413 (1983). The California State Water Resources Control Board was ordered to reconsider its predecessor's grant of water rights to the City of Los Angeles 40 years earlier in light of public trust values, construed by the court to include fish and wildlife habitat and scenic resources.

22 An additional rationale for this restriction is that conservation efforts may reduce return flow and adversely impact other appropriators. But this problem can be guarded against by only allowing consumptive use rights to be bought and sold (see Johnson et al. 1981).

23 For some uses, known as true preferences, no compensation need be paid upon condemnation (see Trelease 1955, 133).

24 For a similar example, see Meyers and Posner (1971, 9–14).

25 This creates another potential political problem if the two states dispute who has first claim to water. Ultimately such disputes can be resolved with interstate compacts, but these are very costly to negotiate.

26 For an in-depth discussion of the incentives of irrigation districts in relation to water transfers, see Thompson 1993, 724–39. Among others, he discusses: (1) the law governing irrigations and conservancy districts often do not allow them to realize a profit from operations; (2) restricting the market for any surplus water to district members limits competition; (3) district managers often want to avoid the controversy that surrounds external transfers; (4) managers fear additional administrative costs generated by external transfers; and (5) managers do not want to reduce the district's water supply because it would reduce their political clout.

27 Under the bureau's 1988 principles, when reclamation water is transferred from irrigation to municipal and industrial uses, transferees are required to pay interest on the repayment obligation assumed from the transferors (Gray et al. 1991, 931).

28 For a detailed discussion of the ways return flow may be altered to the detriment of other water users, see Gould 1988, 13–19. Gould also criticizes redefinition of water rights in terms of consumptive use rather than diversion as tremendously complex, expensive, time-consuming, and wasteful because most water rights will never be transferred. Moreover, redefinition would fail to internalize all externalities of transfers (Gould 1988, 25–28).

29 *Farmers Highline Canal & Reservoir Co. v. City of Golden*, 129 Colo. 575, 272 P.2d 629 (1954).

30   Oregon Revised Statutes §§540.523(2), 540.523(5).
31   The remaining water was picked up by the State Water Project (Hayward 1992, 24).
32   In contrast, transfers of water out of a district or mutual affect the organization as a whole because of conveyance losses and loss of return flows. Such transfers are subject to state transfer procedures since they affect the organization's water rights. Often, they are also either prohibited or opposed by the organization.

6   The New Frontier

1   Chapter 5 details several institutional reforms that would further unlock the potential of markets to avert the water crisis in prior appropriation states, namely relaxing certain restrictions imposed on water uses and transfers.
2   *Stratton v. Mount Hermon Boys School*, 216 Mass. 83, 103 N.E. 87, 89 (1913).
3   See *Mason v. Hoyle*, 14 A. 786, 791 (Conn. 1888) confirming a riparian owner's right to make productive use of water, but not to render unproductive the rights of downstream users); and *Harris v. Brooks*, 283 S.W.2d 129, 136 (Ark, 1955) holding that riparian owner may not exercise his water right in a way detrimental to other users.
4   Restatement (Second) of Torts §850A cmt. e (1979).
5   See, e.g., *Jones v. Oz-Ark-Val Poultry Co.*, 306 S.W.2d 111, 115 (Ark. 1957); *Bouris v. Largent*, 236 N.E2d 15, 17 (Ill. App. Ct. 1968).
6   See, e.g., *Connecticut v. Massachusetts*, 282 U.S. 660 (1931); *Pyle v. Gilbert*, 245 GA 403, 265 S.E.2d 584 (1980).
7   *Hudson River Fishermen's Assn. v. Williams*, 139 AD2d 234, 531 N.Y.S.2d 379 (N.Y.A.D. 3 Dept. 1988).
8   See *Pyle v. Gilbert*, 265 S.E.2d 584, 589 (Ga. 1980); *Mianus Realty Co. v. Greenway*, 193 A.2d 713 (Conn. 1963); *Belvedere Dev. Corp. v. Dep't of Transportation*, 476 So.2d 649 (Fla. 1985).
9   Choe (2004) at 1924, citing *Williams v. Wadsworth*, 51 Conn. 277, at 304.
10   Although the American Society of Civil Engineers' Regulated Riparian Model Water Code incorporates several common principles of regulated riparianism, most states have only adopted small portions of the Model Code. Consequently, state laws differ in terms of the type, volume, and source of water use requiring a permit; the procedure and criteria for issuing a permit; as well as the permit duration and renewal process.
11   The common law riparian doctrine only apportions disputed water resources after a conflict emerges, and the resolutions which are based on the subjective and dynamic considerations of reasonableness can be unstable and unpredictable. Notwithstanding the prohibition on non-riparian use and interbasin transfers, the absence of clearly defined private water rights makes water trading under the common law riparian doctrine practically impossible.
12   Regulated riparianism should be viewed as a piecemeal approach to water allocation because large categories of water users are exempt from the permitting requirement (Sherk 1990, 287, 290–91). In most regulated riparian states, the minimum withdrawal requiring a permit is 100,000 gallons per day—a rather significant amount of water. Given the large number of withdrawals below the exemption amount and their collective impact on any given water course, the permitting process of many states addresses only a fraction of the total water consumption.
13   Dellapenna (1991) indicates that interbasin transfers are allowed in Connecticut, Georgia, Florida, Kentucky, and New York.
14   Fla. Stat. Ann. §373.223(1)(b).
15   It was for similar reasons that western water law did not recognize instream flows as beneficial uses (see Anderson and Johnson 1983).
16   Of course, as Coase (1960) pointed out, once rights are clarified, bargaining will allocate them to their highest valued use if transaction costs are sufficiently low.
17   Note this is precisely the problem with land held under "individual trust" tenure on Indian reservations. Such lands often have multiple, even hundreds, of owners, each of whom can veto land use changes (see Anderson and Lueck 1992).

18   See generally Australian Water Act 2007 (Act no. 137 of 2007, as amended).

19   Based on historical precipitation data, the Compact's drafters expected the 7.5 million acre-feet per year requirement to produce a roughly equal division of the Colorado River water between the Upper and Lower Basin states (Adler 2008). They allowed for smoothing of low-flow and high-flow years by extending the time frame to 10 years and multiplying the annual volumetric requirement by ten.

20   Consideration of instream values goes beyond transportation to include environmental values. Once rights are defined, markets can augment instream flows (see Scarborough 2010), but because instream flows have public good characteristics, a case can be made for mandating minimum flows from the outset (see Chapter 7).

7   Buy That Fish a Drink

1   This chapter primarily focuses on western institutions, but the issue is gaining significance in the East where scarcity and dewatered streams are becoming more common.

2   *Lake Shore Duck Club v. Lake View Duck Club*, 50 Utah 76, 166 P. 309, 310–11 (1917).

3   *Colorado River Water Conservation District v. Rocky Mountain Power Company*, 158 Colo. 331, 406 P.2d 798, 800 (1965).

4   The classic economics example of this element of a public good is the lighthouse. As the argument goes, once a lighthouse is built and lighting the way for passing ships, the additional cost of guiding another ship is zero. Thus, lighthouses must be public goods. But as Ronald Coase (1974) has pointed out, private lighthouses have been in existence for many years and innovative contractual arrangements have evolved to allow private entrepreneurs the return necessary to produce the service.

5   New Mexico laws do not explicitly recognize instream flows as beneficial; however, in 1998 the state's attorney general issued an opinion declaring that the state engineer can protect instream flows for "recreational, fish or wildlife, or ecological purposes" (98-01 at 1 [March 27, 1998]).

6   *McClellan v. Jantzen*, 547 P. 2d 494 (1976).

7   Ariz. Rev. Stat. §45-172.

8   Ariz. Rev. Stat. §45-2101.

9   Cal. Water Code §1707 (a) (1).

10   CA Water Code 1435 provides that any permittee or licensee who has "an urgent need to change a point of diversion, place of use, or purpose of use . . . may petition for . . . a conditional, temporary change order without complying with other procedures or provisions . . ." A temporary urgency change may last for no more than 180 days, but may be renewed by the Board.

11   Colorado Revised Statutes §37-92-102(3)

12   Wood River Minimum Stream Flow, S 1136, Fifty-ninth Legislature, 2007 Session.

13   The Department of Natural Resources and Conservation was formerly called the Board of Natural Resources and Conservation.

14   Reservations must be reviewed at least once every 10 years to ensure that the objectives of the reservation are being met. Reservations may be modified to reallocate all or a portion of the reserved water to another use if the board finds that all or a portion of the reservation is not needed for its designated purpose, and if the other use outweighs the need for the original reservation. Montana Code Annotated 85-2-316(10) (1993).

15   New Mexico Attorney General Opinion 98-01 at 1 (March 27, 1998).

16   Oregon Rev. Stat. 537.336. See Neuman and Chapman (1999) for a thorough description of the statute.

17   Senate Bill 140, The Instream Water Rights Act (1987).

18   Ch. 90.22 RCW.

19   This is a one-time fee, roughly $2,000 per residence well, that covers the cost of the water and associated transaction costs.

20   Wyo. Stat. §41-3-1001(a).

21   Wyo. Stat. §41-3-1002(e).

22   All instream flow acquisition costs and prices have been adjusted to reflect 2007 dollars using the Western Urban Consumer Price Index.

8   Good to the Last Drop

1   The Clean Water Act 502(14), 33 U.S.C. 1362 (14) defines point sources of pollution as any discernible, confined, and discrete conveyance from which pollutants are or may be discharged, including any pipe, ditch, channel, or conduit. Typical point sources include factories, sewage plants and other water treatment facilities, and power plants. Nonpoint source pollution is any source of water pollution not associated with a discrete conveyance (Rodgers 1994, 292), and includes runoff from irrigated lands, stormwater runoff from streets, and sewage overflows.

2   Federal Water Pollution Control Act, 33 U.S.C. 1251–1387.

3   Much of the following material is drawn from the work of Meiners and Yandle (1994a and 1994b).

4   Historically, however, trespass plaintiffs were only required to prove the invasion itself. Intent was irrelevant, and proof of damages was not required.

5   *Maddox v. International Paper Co.*, 47 F. Supp. 829 (W.D. Louisiana 1942).

6   34 Misc. 459, 70 N.Y.S. 284, aff'd 74 N.Y.S. 1145, 175 N.Y. 346, 67 N.E. 622 (1903).

7   200 U.S. 496 (1906).

8   Significantly, the Court stated in its opinion that while the evidence as presented was insufficient to support Missouri's allegations, what the future might show it could not tell. In other words, the Court acknowledged that improved science might enable Missouri to prove its case (*Missouri v. Illinois*, 200 U.S. at 526, 26 S.Ct. at 272).

9   *State of New Jersey v. Ventron Corporation*, 94 N.J. 473, 468 A. 2d 150 (1983). The polluting activities occurred prior to passage of the 1972 CWA and imposition of its permit requirements.

10   In *Milwaukee v. Illinois*, 451 U.S. 304, 101 S.Ct. 1784 (1981) (discussed below), Chief Justice Rehnquist stated:

> The invocation of federal common law by the District Court and the Court of Appeals in the face of congressional legislation supplanting it [the CWA] is peculiarly inappropriate in areas as complex as water pollution control. . . . Not only are the technical problems difficult . . . but the general area is particularly unsuited to the approach inevitable under a regime of federal common law. Congress criticized past approaches to water pollution control as being "sporadic" and "ad hoc," apt characterizations of any judicial approach applying federal common law.
>
> (451 U.S. at 325, 101 S.Ct. 1796–97)

Note, however, that in *Milwaukee*, the chief justice addressed only federal, as opposed to state, common law. Federal common law is created by federal courts in cases where no clear federal statutory or constitutional rule applies, and where the federal courts are not required to apply state law. It is a limited area of law when compared with state common law, which is often well developed but not always clear.

11   For centuries, nuisance law was a body of strict rules that protected landowners' rights to be free from interference with the use and enjoyment of their property. But those rules were loosened in the nineteenth century to accommodate the development that accompanied the Industrial Revolution (Lewin 1990). As of the 1960s, courts had forsaken strict protection of property rights from pollution in favor of balancing the benefit of polluters' conduct to society with individuals' right to be free from unhindered enjoyment of their property. When the polluter was a factory that provided jobs to hundreds of workers and the complaining neighbor was a small farmer whose crops or livestock were being harmed by the pollution, many courts allowed polluters to continue their operations. Although courts often required polluters to pay damages, they did not

enjoin the pollution. Not only was pollution allowed, this balancing act plunged nuisance law into a state of relativity and confusion, leading scholars to describe it as an "impenetrable jungle" (see Prosser and Keeton 1984, 616).

12 Clean Water Act §101(a), 33 U.S.C. 1251(a) (1988). The CWA addresses surface water only; groundwater is not covered.

13 Clean Water Act §101(a)(1) & (2), 33 U.S.C. 1251(a)(1) & (2) (1988).

14 Apart from grant money, the states see combined sewer overflow controls as unfunded mandates, i.e., federally imposed requirements on local governments without appropriating money to pay for them (Hanson 1994, 38).

15 Under the CWA, states are authorized to administer their own discharge permit programs subject to the EPA's approval of each state's program. The CWA also authorizes states to issue water quality standards so long as they are consistent with the act. Management of water quality on a watershed basis would shake up this established system of state authority by removing it from the states and giving it to local and regional basin authorities subject to the EPA.

16 For a discussion of how agricultural subsidies encourage wetland drainage, see Gardner (1995).

17 Wetlands have generated considerable controversy. In 2006, the Supreme Court's decision in *Rapanos v. United States* (U.S. No. 04-1034) challenged the jurisdiction of the U.S. Army Corps and the EPA in its ability to regulate wetlands that were not clearly adjacent to "navigable waters." The Supreme Court held that isolated wetlands were not considered "waters of the United States" and thus not subject to regulations under the CWA (see also *Carabell v. United States* U.S. No. 04-1034). As discussed in Chapter 4, the Clean Water Restoration Act of 2009, if passed, would broaden the regulatory reach of the CWA to include all "waters of the United States," including isolated wetlands not adjacent to navigable waters.

18 Hearing on Governors' Perspectives on the Clean Water Act Before the Subcommittee on Water Resources and Environment of the Committee on Transportation and Infrastructure, House of Representatives, One Hundred Sixth Congress, first session, February 23, 1999 (quoted in Houck 2002, 139).

19 451 U.S. 304, 101 S.Ct. 1784 (1981).

20 479 U.S. 481, 107 S.Ct. 805 (1987).

21 Although under *Ouellette* state common law survived preemption by the CWA at least in part, its viability is limited. First, state common law will survive preemption only if it does not stand as an obstacle to the full implementation of the CWA. Second, as a practical matter, proving that a discharger should be liable for damages under the common law when it is complying with a federal water pollution law can be a challenge.

22 Clean Water Act §505(a), 33 U.S.C. §1365(a). The citizen suit provision applies only to point source dischargers because effluent limitations and permits are not applicable to nonpoint sources.

23 In fact, permit holders have successfully raised what has become known as the "permit as a shield" defense, which entails using their permits as a defense against liability for discharging a pollutant not covered by the permit. On December 14, 1994, the "Second Circuit Court of Appeals held that as long as a discharger complies with effluent limits established for the specifically listed pollutants in its permit, then it is shielded from enforcement actions for discharging any pollutant among the universe of pollutants not regulated in the permit" (*U.S. Water News* 1994a, 1). Not all federal courts agree with the Second Circuit's decision, however. The permit as a shield defense is debated within the federal court system and among government and private environmental attorneys (*U.S. Water News* 1994a, 1).

24 *Gwaltney of Smithfield, Ltd. v. Chesapeake Bay Foundation*, 484 U.S. 49 (1987).

25 Order dated February 8, 1995, in *Knee Deep Cattle Co., et al. v. Bindana Investments Co., et al.*, Civil No. 94-6156-TC, U.S. District Court, District of Oregon. Much of the following information came from Moon (1995).

26 Findings and Recommendation of the U.S. Magistrate Judge in *Knee Deep Cattle Co. v. Bindana Investments Co.*, 13 (citing *Saboe v. Oregon*, 819 F. Supp. 914 [D. Ore. 1993]).

27 *Knee Deep Cattle Co. v. Bindana Inv. Co.*, 94 F.3d 514 (9th Cir. 1996).

28 It is important to note that most discussions of water quality markets focus on one pollutant at a time or closely related pollutants such as nitrogen and phosphorus, which are both nutrients, as opposed to addressing all pollutants that enter a waterbody. The pollution problems suffered by many basins generally result from excessive amounts of either nonconventional (typical of municipal sewage) or conventional pollutants (such as phosphorus and nitrogen). Because they decay over time, control of nonconventional pollutants such as biochemical oxygen demand (BOD) requires staggering discharges by time and location so that pollution levels do not violate water quality standards. A major source of BOD is publicly owned sewage treatment works. In contrast, conventional pollutants cause trouble when they accumulate and tend to be a problem in waterbodies that are not regularly flushed out. Concentrations of conventional materials do not change as water moves downstream because such materials do not degrade or decay. According to Letson (1992), trading programs are simpler and more appealing when they involve conventional pollutants, as opposed to nonconventional pollutants which are uncertain and are a politically charged issue (Letson 1992, 226; Bartfeld 1993, 63).

29 The task of directly monitoring individual dischargers would become virtually impossible if nonpoint sources were added to the regulatory burden of states and the EPA under the CWA.

30 In a worldwide survey, the World Resources Institute identified six trading programs outside the United States—four active programs (three in Australia and one in New Zealand) and two in development (one in Australia and one in New Zealand) (Selman et al. 2009).

31 There is sufficient flexibility in the program to provide for trades with nonpoint sources. However, the costs of nonpoint source controls have proven to be higher than point source controls (Breetz et al. 2004). The acquisition of point source credits could be used as a cost-effective means of offsetting nonpoint source targets in the future.

32 To be clear, robust trading activity is not always necessary to accomplish water quality targets. In some instances, few but large isolated trades may be sufficient to offset excess point source discharges. For example, the creation of riparian buffers or tree planting programs as in the Tualatin River basin in Oregon, can provide long-term credits for up to 30 years, meaning the point source acquirer of those credits may not need to buy additional credits unless future changes in regulatory caps require him to do so.

33 There is still some uncertainty surrounding where the liability falls in cases where nutrient reduction projects fail to provide actual reductions in loadings. By default, the buyer assumes liability because of permit requirements that are enforced by the permitting authority. In some cases, a third-party broker, exchange or clearinghouse may certify tradable credits and thus assume some of the risks. However, it is still the regulated point source that must meet loading targets. To reduce risk to buyers, a clearinghouse or exchange may bank extra credits to ensure sufficient reductions can be made even in the event of some nutrient management projects falling short of their targets.

9 The Race to Pump

1 Public Law 94-587 (Sec. 193) 94th Congress.

2 According to Heath (1988, 76–77), permeability, or hydraulic conductivity, is a function of pore size and determines the water-transmitting capacity of the rock. Porosity is a function of the number of pores in the rock, and determines its storage capacity.

3 Well interference may also be nonexistent in highly transmissive aquifers, such as the Edwards Aquifer in Texas, where water moves quickly through the aquifer to replace water that is withdrawn. In such aquifers, pumping results in an aquifer-wide drop in the water table.

4   12 M. and W. 349, 152 Eng. Rep. 1223 (Exch. 1843).

5   12 M. and W. at 349, 152 Eng. Rep. at 1223.

6   525 Ill. Comp. Stat. 45/4, (1983) (emphasis added).

7   Ga. Comp. R. & Regs. 391-3-2 (2004).

8   The Restatement (Second) of Torts, Section 858 states:

> (1) A proprietor of land or his grantee who withdraws ground water from the land and uses it for a beneficial purpose is not subject to liability for interference with the use of the water by another, unless (a) the withdrawal of ground water unreasonably causes harm to a proprietor of neighboring land through lowering the water table or reducing artesian pressure, (b) the withdrawal of ground water exceeds the proprietor's reasonable share of the annual supply or total store of ground water, or (c) the withdrawal of groundwater has a direct and substantial effect upon a watercourse or lake and unreasonably causes harm to a person entitled to the use of its water.

> (2) The determination of liability under clauses (a), (b) and (c) of Subsection (1) is governed by the principles stated in sections 850 to 857 [which set forth the Restatement rules governing riparian water rights, i.e., reasonable use].

9   The groundwater permit schemes in states following the reasonable use or modified reasonable use rule typically do not distinguish between groundwater uses during times of average or surplus groundwater.

10 *Katz v. Walkinshaw* (1903) 141 Cal. 116 [74 P. 766, 722].

11 *City of Barstow v. Mojave Water Agency* (2000) 23 Cal. 4th 1224 [5 P.3d 853].

12 A basin is in a state of "surplus" when the amount of water being extracted from it is less than the maximum amount that could be drawn without adverse effects on the basin's long-term supply.

13 33 Cal. 2d 908, 207 P.2d 17 (1949).

14 As the California Supreme Court later clarified in *City of Barstow v. Mojave Water Agency* (2000) 23 Cal. 4th 1224, 1241, these prescriptive rights arose from the appropriators' actual, open, notorious, adverse, exclusive and continuous use of water to which someone else—the overlying landowners—held a right.

15 14 Cal. 3d 199, 537 P.3d 1250, 123 Cal Reptr. 1 (1975).

16 *City of Barstow v. Mojave Water Agency* (2000), 23 Cal. 4th 1224 [5 P.3d 853].

17 Conservation organizations and senior rights holders have recently attempted to eliminate the state's permit exemption for wells pumping less than 10 acre-feet per year at no more than 35 gallons per minute. These small, often single-family, wells were once thought inconsequential to the overall function of groundwater aquifers. But as population density and groundwater pumping increased, these exempt wells have cumulatively reduced the amount of groundwater available for senior appropriators and instream uses. In response to this scarcity, Montana's groundwater policy will likely change again.

18 For a more complete discussion, see Provencher and Burt (1993).

19 This analysis is drawn largely from Tehachapi Soil Conservation District (1969), Gates (1969), and Lipson (1978), as well as conversations with John Otto, assistant manager, Tehachapi–Cummings County Water District, Tehachapi, California. For a more detailed discussion, see Anderson et al. (1983).

20 Justifications for this rule are that excessive stockpiling might cause the aquifer to fill up, thus impairing surface drainage; and if too many stockpiled rights are exercised at one time, there might not be enough water to satisfy all demands due to cones of depression.

21 Note that the average pumping costs for groundwater are determined by the watermaster and applied to all participants in the exchange pool regardless of their actual pumping costs.

22 179 Okla. 53, 64 P.2d 694 (1937).

23 According to Roberts and Gros (1987), the 1949 law failed for several reasons. First, the total conservation policy prohibiting basin overdraft was untenable because it drastically limited agricultural activity throughout much of the state. Second, pumpers assumed groundwater permits would be routinely granted according to property rights, so pumping continued unabated. Rather than curtail the pumping, the Division of Water Resources arbitrarily issued permits of 2 acre-feet per acre per year. And finally, there was a widespread lack of concern over groundwater depletion, so there was widespread apathy towards this conservation measure.

24 Okla. Stat. tit. 82 ch. 11 §1020.5 (1973).

10 You Can't Keep a Good Market Down

1 Las Vegas, Nevada Code of Ordinances 14.08.030; see also Ord. 247 §1, 1943: Ord. 218 §1, 1936: Ord. 211 §1, 1934: prior code §3-2-4.

2 Xeriscaping is a landscaping or gardening technique that employs the use of water-tolerant plants, reducing the need for supplemental irrigation and water use.

3 Idaho Administrative Rules 37.02.04 (1994).

4 Decree Law 2,603, passed in 1979, was an important precursor to the 1981 Water Code because it created the legal separation of water and land rights, and allowed for private water trading. Decree Law 2,603 also created a tax on water rights similar to a real estate tax, but this provision was eventually omitted from the 1981 Water Code.

5 Under the 1981 Code, the DGA had no discretion to consider environmental impacts. In 2005, the Water Code was reformed and the DGA was granted authority to consider minimum stream flows and sustainable aquifer management in granting new water rights (Williams and Carriger 2005).

6 The 1981 Water Code validated rights "recognized by executive rulings . . . those arising from grants given by competent authority . . . and those acquired by prescription" (Donoso 2006).

7 This taxonomy distinguishes between consumptive and non-consumptive uses, permanent or contingent, and continuous or discontinuous. Consumptive uses do not require that the water be returned after use, whereas non-consumptive use rights require that the water be returned to the water course in a manner that does not interfere with consumptive uses. Permanent rights are rights to use water in specified amounts, while contingent rights only authorize the user to utilize water when the flow is sufficient to satisfy all permanent rights. Continuous rights permit withdrawal at any time of day, all year long; while discontinuous rights have specified periods of withdrawal (Donoso 2006).

REFERENCES

1 Why the Crisis?

Alley, William M., Richard W. Healy, James W. LaBaugh, and Thomas E. Reilly. 2002. Flow and Storage in Groundwater Systems. *Science* 14 (5575): 1985–90.

Barrett, Joe. 2009. City's Water Problems Test Great Lakes Agreement. *Wall Street Journal*, November 9.

Barlow, Maude. 2007. *Blue Covenant*. New York: The New Press.

Berfield, Susan. 2008. There Will Be Water. *Business Week*, June 12.

Cosgrove, W. J., and F. R. Rijsberman. 2000. *World Water Vision: Making Water Everybody's Business*. London: Earthscan Publications Ltd.

Craft, Elena, S. Kirby, C. Donnelly, Iulia Neamtiu, Kathleen M. McCarty, Erica Bruce, Irina Surkova, David Kim, Iveta Uhnakova, Erika Gyorffy, Eva Tesarova, and Beth Anderson. 2006. Prioritizing Environmental Issues Around the World: Opinions from an International Central and Eastern Europeans Environmental Health Conference. *Environmental Health Perspectives* 114 (12): 1813–17.

Dosi, Cesare, and K. William Easter. 2003. Market Failure and Role of Markets and Privatization in Alleviating Water Scarcity. *International Journal of Public Administration* 26 (3): 265–90.

Gleick, Peter H., Heather Cooley, Michael Cohen, Mari Morikawa, James Morrison, and Meena Palaniappan. 2009. *The World's Water 2008–2009: The Biennial Report on Freshwater Resources*. Washington, DC: Island Press.

Gleick, Peter H., Heather Cooley, David Katz, Emily Lee, Jason Morrison, Meena Palaniappan, Andrea Samulon, and Gary Wolff. 2006. *The World's Water 2006–2007: The Biennial Report on Freshwater Resources*. Washington, DC: Island Press.

Glennon, Robert J. 2002. *Water Follies: Groundwater Pumping and The Fate of America's Fresh Waters*. Washington, DC: Island Press.

Jansen, John, and Steven Schultz. 2009. Science, Politics, and the Future Water Supply for Southeastern Wisconsin. Article Five, Aquifer Science and Technology. www.aquiferscience.com/Article5.htm (accessed: November 18, 2009).

Lomborg, Bjørn. 2001. *The Skeptical Environmentalist: Measuring the Real State of the World*. Cambridge: Cambridge University Press.

Mackun, Paul, and Steven Wilson. 2011. Population Distribution and Change: 2000 to 2010. 2010 Census Briefs, C2010BR-01, March. www.census.gov/prod/cen2010/briefs/c2010br-01.pdf (accessed: May 12, 2011).

McGuire, V. L., M. R. Johnson, R. L. Schieffer, J. S. Stanton, S. K. Sebree, and I. M. Verstraeten. 2003. Water in Storage and Approaches to Ground-water Management, High Plains Aquifer, 2000. United States Geological Survey Circular, no. 1243.

Pearce, Fred. 2006. *When the Rivers Run Dry: Water—The Defining Crisis of the Twenty-First Century*. Boston: Beacon Press Books.

Sachs, Jeffrey D. 2008. *Common Wealth: Economics for a Crowded Planet*. New York: Penguin Press.

United Nations Development Programme (UNDP). 2006. *Beyond Scarcity: Power, Poverty and the Global Water Crisis*. New York: United Nations Development Programme.

United Nations Environment Programme (UNEP). 2007. *Global Environment Outlook: Environment for Development (GEO-4)*. Valletta, Malta: Progress Press Ltd. www.unep.org/geo/geo4.asp (accessed: May 12, 2011).

——. 2008. *Vital Water Graphics: An Overview of the State of the World's Fresh and Marine Waters*, 2nd ed. UNEP/GRID-Arendal. www.unep.org/dewa/vitalwater/article186.html (accessed: May 12, 2011).

U.S. Census Bureau, Population Division. 2009. Annual Estimates of the Population of Combined Statistical Areas: April 1, 2000 to July 1, 2008. www.census.gov/population/estimates/metro_general/2008/CSA-EST2008-alldata.csv (accessed: April 21, 2009).

U.S. Environmental Protection Agency (U.S. EPA) and the Government of Canada. 2002. *The Great Lakes: An Environmental Atlas and Resource Book*. Chicago, Illinois, and Toronto, Canada.

U.S. Geological Survey (USGS). 2009. The Water Cycle: Freshwater Storage. ga.water.usgs.gov/edu/watercyclefreshstorage.html (accessed: May 13, 2009).

U.S. Water Resources Council. 1978. *The Nation's Water Resources, 1975–2000*. Washington, DC: Government Printing Office.

Western Water Assessment. 2011. "The Lees Ferry Gauged Flow Record." *Colorado River Streamflow: A Paleo Perspective*. http://wwa.colorado.edu/treeflow/lees/gage.html (accessed: May 12, 2011).

World Bank. 2003. *World Development Report 2003: Sustainable Development in a Dynamic World: Transforming Institutions, Growth, and Quality of Life*. New York: Oxford University Press.

World Health Organization (WHO) and United Nations Children's Fund. 2000. *Global Water Supply and Sanitation Assessment 2000 Report*. www.who.int/docstore/water_sanitation_health/Globassessment/GlobalTOC.htm (accessed: May 12, 2011).

World Health Organization and United Nations Children's Fund Joint Monitoring Programme for Water Supply and Sanitation (JMP). 2008. *Progress on Drinking Water and Sanitation: Special Focus on Sanitation*. UNICEF, New York and WHO, Geneva.

World Resources Institute. 2008. Coastal Populations Losing Livelihoods to Polluted Water. www.wri.org/press/2008/04/coastal-populations-losing-livelihoods-polluted-waters (accessed: May 12, 2011).

World Water Council. 2006. Final Report of the 4th World Water Forum. Mexico City: National Water Commission of Mexico. www.worldwatercouncil.org/fileadmin/wwc/Library/ Publications_and_reports/ Final_Report_4th_Forum.pdf. (accessed: July 12, 2009).

Zimmerman, Julie Beth, James R. Mihelcic, and James Smith. 2008. Global Stressors on Water Quality and Quantity. *Environment Science and Technology* 42 (12): 4247–54.

2 Cheaper Than Dirt

Anderson, Terry L. 2004. *You Have to Admit It's Getting Better: From Economic Prosperity to Environmental Quality*. Stanford, CA: Hoover Institution Press.

Anderson, Terry L., and Peter J. Hill. 1975. The Evolution of Property Rights: A Study of the American West. *Journal of Law and Economics* 18 (1): 163–79.

———. 1980. *The Birth of a Transfer Society*. Stanford, CA: Hoover Institution Press, 1980.

Anderson, Terry L., and Robert E. McCormick. 2004. The Contractual Nature of the Environment. In *The Elgar Companion to Property Rights*, ed. Enrico Colombatto. New York: Edward Elgar, 293–309.

Bailey, Ronald, ed. 1995. *The True State of the Planet*. New York: Free Press.

Barlow, Maude. 2007. *Blue Covenant: The Global Water Crisis and the Coming Battle for the Right to Water*. New York: The New Press.

Barney, Gerald O. 1980. *The Global 2000 Report to the President. Volume I: Entering the Twenty-First Century*. New York: Pergamon Press.

Caswell, Margriet, and David Zilberman. 1985. The Choice of Irrigation Technologies in California. *American Journal of Agricultural Economics* 67 (2, May): 224–34.

Coase, Ronald H. 1960. The Problem of Social Cost. *Journal of Law and Economics* 3: 1–44.

Demsetz, Harold. 1967. Toward a Theory of Property Rights. *American Economic Review* 57: 347–59.

———. 2003. Ownership and the Externality Problem. In *Property Rights: Cooperation, Conflict, and Law*, ed. Terry L. Anderson and Fred S. McChesney. Princeton, NJ: Princeton University Press.

Diamond, Jared M. 2005. *Collapse: How Societies Choose to Fail or Survive*. New York: Viking Press.

Ehrlich, Paul R. 1968. *The Population Bomb*. New York: Ballantine Books.

Espey, M., J. Espey, and W. D. Shaw. 2007. Price Elasticity of Residential Demand for Water: A Meta-Analysis. *Water Resources Research* 33 (6): 1369–74.

Glennon, Robert. 2009. *Unquenchable: America's Water Crisis and What to Do About It*. Washington, DC: Island Press.

Gould, George A. 1988. Water Rights Transfers and Third-Party Effects. *Land and Water Law Review* 23: 1–41.

Hardin, Garrett. 1968. The Tragedy of the Commons. *Science* 162 (3859): 1243–48.

Hayek, Friedrich A. 1945. The Use of Knowledge in Society. *American Economic Review* 35 (4): 519–30.

Hays, Samuel P. 1959. *Conservation and the Gospel of Efficiency: A Progressive Conservation Movement, 1890–1920*. Cambridge, MA: Harvard University Press.

Hedges, Trimble. 1977. *Water Supplies and Costs in Relation to Farm Resource Decisions and Profits on Sacramento Valley Farms*. Report 322. Berkeley, CA: Gianinni Foundation.

Hirshleifer, Jack, James C. DeHaven, and Jerome W. Milliman. 1960. *Water Supply: Economics, Technology, Policy*. Chicago: University of Chicago Press.

Keen, Judy. 2008. Gas Costs Cut into Vacation Travel. *USA Today*, May 23.

Kenny, J. F., N. L. Barber, S. S. Hutson, K. S. Linsey, J. K. Lovelace, and M. A. Maupin. 2009. Estimated Use of Water in the United States in 2005. *U.S. Geological Survey Circular* 1344.

Lomborg, Bjørn. 2001. *The Skeptical Environmentalist: Measuring the Real State of the World*. Cambridge: Cambridge University Press.

McChesney, Fred S. 1997. *Money for Nothing: Politicians, Rent Extraction, and Political Extortion*. Cambridge, MA: Harvard University Press.

Meadows, Donella H., Dennis L. Meadows, Jorgen Randers, and William W. Behrens III. 1972. *The Limits to Growth*. New York: Universe Books.

Mehan, G. Tracy. 2008. Let's Drink to Private Water. *PERC Reports* 26 (1, Spring): 22–25.

Mitchell, William C., and Randy T. Simmons. 1994. *Beyond Politics: Markets, Welfare, and the Failure of Bureaucracy*. Boulder, CO: Westview Press.

Olmstead, Sheila M., and Robert N. Stavins. 2007. Managing Water Demand Price vs. Non-Price Conservation Programs. *Pioneer Institute White Paper* 39: 1–47.

Pigou, A. C. 1932. *The Economics of Welfare*. London: Macmillan and Co.

Reynaud, Arnaud. 2003. An Econometric Estimation of Industrial Water Demand in France. *Environmental and Resource Economics* 25 (2): 213–32.

Sachs, Jeffrey D. 2008. *Common Wealth: Economics for a Crowded Planet*. New York: Penguin Press.

Scarborough, Brandon. 2010. Environmental Water Markets: Restoring Streams Through Trade. *PERC Policy Series* no. 46. Bozeman, MT: PERC.

Scarborough, Brandon, and Hertha Lund. 2007. *Saving Our Streams: Harnessing Water Markets*. Bozeman, MT: PERC.

Scheierling, Susanne M., John B. Loomis, and Robert A. Young. 2006. Irrigation Water Demand: A Meta-Analysis of Price Elasticities. *Water Resources Research* 42: 1–9.

Simon, Julian L., ed. 1995. *The State of Humanity*. Cambridge, MA: Blackwell Publishers.

Tullock, Gordon. 1967. The Welfare Costs of Tariffs, Monopolies, and Theft. *Western Economic Journal* 5 (3): 224–32.

U.S. Department of the Interior. 2009. Budget Justifications and Performance Information Fiscal Year 2009. Central Utah Project Completion Act. www.doi.gov/budget/2009/data/greenbook/FY2009_CUPCA_Greenbook.pdf. (accessed: May 12, 2011).

U.S. Environmental Protection Agency (U.S. EPA). 2002. Clean Water and Drinking Water Infrastructure Gap Analysis. www.epa.gov/waterinfrastructure/infrastructuregap.html. (accessed: May 12, 2011).

U.S. General Accounting Office. 2002. Water Infrastructure: Information on Financing, Capital Planning and Privatization. Report to Congressional Requesters, GAO-02-764.

Water Strategist. Various years. Claremont, CA: Stratecon, Inc.

3 Who Owns the Water?

Alston, Richard Moss. 1978. *Commercial Irrigation Enterprise, The Fear of Water Monopoly, and the Genesis of Market Distortion in the Nineteenth Century American West*. New York: Arno Press.

Anderson, Terry L., and Peter J. Hill. 1975. The Evolution of Property Rights: A Study of the American West. *Journal of Law and Economics* 18 (April): 163–79.

———. 1990. The Race for Property Rights. *Journal of Law and Economics* 33 (1): 177–97.

———. 1995. *Wildlife in the Marketplace*. Lanham, MD: Rowman and Littlefield.

———. 2004. *The Not So Wild, Wild West: Property Rights on the Frontier*. Stanford, CA: Stanford University Press.

Anderson, Terry L., and Donald R. Leal. 2001. *Free Market Environmentalism*. New York: Palgrave.

Baker, Donald M., and Harold Conkling, 1930. *Water Supply and Utilization*. New York: John Wiley & Sons.

Bretsen, Stephen N., and Peter J. Hill. 2007. Irrigation Institutions in the American West. *UCLA Journal of Environmental Law and Policy* 25 (2): 283–331.

Caponera, Dante Augusto. 2007. *Principles of Water Law and Administration*. London: Taylor and Francis.

Chandler, A. E. 1913. *Elements of Western Water Law.* San Francisco: Technical Publishing Company.

Dellapenna, Joseph W. 1991. The Legal Regulation of Diffused Surface Water. *Villanova Environmental Law Journal* 2: 285–332.

———, ed. 1997. *The Regulated Riparian Model Water Code.* New York: American Society of Civil Engineers. Updated in American Society of Civil Engineers, *The Regulated Riparian Model Water Code.* Reston: American Society of Civil Engineers.

———. 2001. The Case against Markets. In *UNESCO International Hydrological Programme,* IHP-V, ed. Uri Shamir. Technical Documents in Hydrology, no. 53.

———. 2002. The Law of Water Allocation in the Southeastern States at the Opening of the Twenty-First Century. *University of Arkansas at Little Rock Law Review* 25: 9–88.

———. 2004. Special Challenges to Water Markets in Riparian States. *Georgia State University Law Review* 21 (2): 305–22.

Demsetz, Harold. 1967. Toward a Theory of Property Rights. *American Economic Review* 57: 347–59.

Dunbar, Robert G. 1983. *Forging New Rights in Western Waters.* Lincoln: University of Nebraska Press.

Golze, Alfred R. 1961. *Reclamation in the United States.* Caldwell, ID: The Caxton Printers.

Gordon, H. S. 1954. The Economic Theory of a Common-Property Resource: The Fishery. *Journal of Political Economy* 62: 124–42.

Huffman, Roy E. 1953. *Irrigation Development and Public Water Policy.* New York: The Ronald Press Company.

Hutchins, Wells A., H. E. Selby, and Stanley W. Voelker. 1953. Irrigation-Enterprise Organizations. *United States Department of Agriculture Circular No. 934.* Washington, DC: United States Government Printing Office.

Kinney, Clesson. 1912. *Law of Irrigation and Water Rights and Arid Region Doctrine of Appropriation of Waters, Vol. 1.* San Francisco: Bender-Moss.

Leal, Donald R. 2004. *Evolving Property Rights in Marine Fisheries.* Lanham, MD: Rowman and Littlefield.

Leshy, John D. 1982. Irrigation Districts in a Changing West: An Overview. *Arizona State Law Journal* 2: 345–76.

Lueck, Dean. 2003. First Possession as the Basis of Property. In *Property Rights: Cooperation, Conflict, and Law,* ed. Terry L. Anderson and Fred S. McChesney. Princeton, NJ: Princeton University Press, 200–26.

Marcus, Richard R., and Stephen Kiebzak. 2008. The Role of Water Doctrines in Enhancing Opportunities for Sustainable Agriculture in Alabama. *Journal of the American Water Resources Association* 44 (6): 1578–590.

McCurdy, Charles W. 1976. Stephen J. Field and Public Land Law Development in California, 1850–1866: A Case Study of Judicial Resource Allocation in Nineteenth-Century America. *Law and Society Review* 10 (1): 235–66.

Mead, Elwood. 1903. *Irrigation Institutions.* New York: Macmillan.

Miano, Steven T., and Michael E. Crane. 2003. Eastern Water Law: Historical Perspectives and Emerging Trends. *Natural Resources and Environment* 18 (2): 129–33.

Pisani, Donald J. 1992. *To Reclaim A Divided West: Water, Law, and Public Policy, 1848–1902.* Albuquerque: University of New Mexico Press.

Smith, Rodney T. 1984. *Troubled Waters: Financing Water in the West.* Washington: Council of State Planning Agencies.

Tarlock, Dan A., James N. Corbridge, and David H. Getches. 2002. *Water Resource Management.* New York: Foundation Press.

Teclaff, L.A. 1985. *Water Law in Historical Perspective*. Buffalo, NY: William S. Hein Company.

Webb, Walter Prescott. 1931. *The Great Plains*. New York: Grosset and Dunlap.

Worster, Donald. 1985. *Rivers of Empire: Water, Aridity and the Growth of the American West*. New York: Pantheon Books.

4   Water Is for Fightin'

Alston, Richard M. 1978. *Commercial Irrigation Enterprise: The Fear of Water Monopoly and the Genesis of Market Distortion in the Nineteenth Century American West*. New York: Arno Press.

Arnold, Craig A. 2004. Working Out Environmental Ethic: Anniversary Lessons From Mono Lake. *Wyoming Law Review* 4 (1): 1–55.

Bain, Joe S., Richard E. Caves, and Julius Margolis. 1963. *Northern California's Water Industry: The Comparative Efficiency of Public Enterprise in Developing a Scarce Natural Resource*. Baltimore, MD: Johns Hopkins Press for Resources for the Future, Inc.

Baughman, John F. 1992. Balancing Commerce, History, and Geography: Defining the Navigable Waters of the United States. *Michigan Law Review* 90 (March): 1028.

Boxall, Bettina. 2011. A Small Fish Caught in a Big Fuss. *Los Angeles Times*, February 2.

California Department of Fish and Game. 2004. September 2002 Klamath River Fish-Kill: Final Analysis of Contributing Factors and Impacts. www.pcffa.org/KlamFishKill FactorsDFGReport.pdf (accessed: May 6, 2011).

Clayberg, John B. 1902. The Genesis and Development of the Law of Waters in the Far West. *Michigan Law Review* 1 (November): 91–101.

Coase, Ronald H. 1960. The Problem of Social Cost. *Journal of Law and Economics* 3: 1–44.

Corker, Charles E. 1957. Water Rights and Federalism: The Western Water Rights Settlement Bill of 1957. *California Law Review* 45: 604–37.

Cuzan, Alfred G. 1983. Appropriators vs. Expropriators: The Political Economy of Water in the West. In *Water Rights: Scarce Resource Allocation, Bureaucracy, and the Environment*, ed. Terry L. Anderson. San Francisco: Pacific Institute for Public Policy Research, 13–43.

Dunbar, Robert G. 1983. *Forging New Rights in Western Waters*. Lincoln: University of Nebraska Press.

Gardner, B. Delworth. 1995. *Plowing Ground in Washington: The Political Economy of U.S. Agriculture*. San Francisco: Pacific Research Institute.

Gardner, B. Delworth, and John E. Warner. 1994. Two Steps Forward—One Step Back. *Choices* (First Quarter): 5–9.

Glennon, Robert. 2009. *Unquenchable: America's Water Crisis and What to Do About It*. Washington, DC: Island Press.

Golze, Alfred R. 1961. *Reclamation in the United States*. Caldwell, ID: The Caxton Printers.

Goplerud, C. Peter, III. 1995. Water Pollution Law: Milestones from the Past and Anticipation of the Future. *Natural Resources and Environment* 10 (Fall): 7–12.

Hartman, L. M. and Don Seastone. 1970. *Water Transfers: Economic Efficiency and Alternative Institutions*. Baltimore, MD: Johns Hopkins University Press.

Hathaway, Ron. 2001. Klamath Water Allocation Background. In *Water Allocation in the Klamath Basin: An Assessment of Natural Resource, Economic, Social, and Institutional Issues*. Draft Report. Oregon State University and the University of California, December.

Hirshleifer, Jack, James C. DeHaven, and Jerome W. Milliman. 1960. *Water Supply: Economics, Technology, Policy*. Chicago: University of Chicago Press.

Howitt, Richard E., Duncan MacEwan, and Josue Medellin-Azuara. 2009. Economic Impacts of Reductions in Delta Exports on Central Valley Agriculture. *Agricultural and Resource Economics Update* 12 (3): 1–4.

Lasky, Moses. 1929. From Prior Appropriation to Economic Distribution of Water by the State—Via Irrigation Administration. *Rocky Mountain Law Review* 1 and 2 (April, June, November): 161–216, 248–270, 35–58.

Libecap, Gary. 2006. The Battle Over Mono Lake. *Hoover Digest* 2: 88–94.

Marston, Ed. 1994. On Friday, the Fish Took Some of It Back. *High Country News*, October 31, 15.

McCurdy, Charles W. 1976. Stephen J. Field and Public Land Law Development in California, 1850–1866: A Case Study of Judicial Resource Allocation in Nineteenth-Century America. *Law and Society Review* 10 (1): 235–66.

McDevitt, Edward. 1994. The Evolution of Irrigation Institutions in California: The Rise of the Irrigation District, 1910–1930. Ph.D. dissertation, Department of Economics, University of California–Los Angeles, Los Angeles, California.

Meyers, Charles J. and Richard A. Posner. 1971. Market Transfers of Water Rights, Legal Study no. 4. Washington, DC: National Water Commission.

Minardi, Jean-Francois. 2010. Going Down the Tubes: Montreal Needs to Fix Its Water Infrastructure—and Fast. *Fraser Forum*, July/August, 8–9.

National Research Council. 2002. *Privatization of Water Services in the United States: An Assessment of Issues and Experience.* Washington, DC: National Academy Press.

Sax, Joseph L., and Robert H. Abrams. 1986. *Legal Control of Water Resources: Cases and Materials.* St. Paul, MN: West Publishing Company.

Segerfeldt, Frederik. 2005. Private Water Saves Lives. *Financial Times*, August 25.

Slade, David C. 2008. *The Public Trust Doctrine in Motion 1997–2008.* Bowie, MD: PTDIM, LLC.

Smith, Rodney T. 1992. Water Rights Claims in Indian Country: From Legal Theory to Economic Reality. In *Property Rights and Indian Economies*, ed. Terry L. Anderson. Lanham, MD: Rowman and Littlefield, 167–94.

Smith, Rodney T., and Roger Vaughan. 1991. Interior's Policy of Voluntary Water Transactions: The Two-Year Record. *Water Strategist* 4 (January): 1–13.

U.S. Bureau of Reclamation. 2010. Weber Siphon Complex—Odessa Subarea. *Reclamation: Managing Water in the West.* http://recovery.doi.gov/press/wp-content/uploads/2010/08/weber-siphon072210.pdfrecovery.doi.gov/press/wp-content/uploads/2010/08/weber-siphon072210.pdf (accessed: May 6, 2011).

U.S. Congressional Budget Office (U.S. CBO). 2006. How Federal Policies Affect the Allocation of Water. CBO paper, publication no. 2589, August. www.cbo.gov/ftpdocs/74xx/doc7471/08-07-WaterAllocation.pdf (accessed: May 6, 2011).

U.S. Dept. of the Interior. 1988. Principles Governing Voluntary Water Transactions That Involve or Affect Facilities Owned or Operated by the Department of Interior. Washington, DC. December 16.

U.S. Fish and Wildlife Service. 1993. Endangered and Threatened Wildlife and Plants; Determination of Threatened Status for the Delta Smelt. *Federal Register* 58 (42, 5 March): 12854–863.

Washburn, Edgar B. 1986. The Implications of the Public Trust Doctrine for Land and Water Titles. In *Western Resources in Transition: The Public Trust Doctrine and Property Rights.* Conference Proceedings, Political Economy Research Center, Bozeman, Montana, May 17.

Washington Department of Ecology. 2006. Washington State Water Right Adjudication Process: A Primer. Publication #WR 98-151. www.ecy.wa.gov/pubs/98151.pdf (accessed: May 6, 2011).

Wilkinson, Charles F. (1991) In Memoriam: Prior Appropriation 1848–1991. *Environmental Law* 21 (3): v.

5   Back to the Future

Anderson, Terry L., and Ronald N. Johnson. 1986. The Problem of Instream Flows. *Economic Inquiry* 24 (4): 535–54.

Bates, Sarah, Charles Wilkinson, Lawrence MacDonnell and David Getches. 1993. *Searching Out the Headwaters: Change and Rediscovery in Western Water Policy*. Albuquerque: University of New Mexico Press.

Brewer, Jedidiah, Robert Glennon, Alan Ker, and Gary Libecap. 2006. Water Markets in the West: Prices, Trading, And Contractual Forms. *Economic Inquiry* 46 (2): 91–112.

Brown, F. Lee. 2007. Beyond the Year of Water: Living Within Our Water Limitations. New Mexico Water Resources Research Institute, 52nd Annual New Mexico Water Conference, November.

Burness, H. Stuart, and James P. Quirk. 1980. Water Laws, Water Transfers, and Economic Efficiency: The Colorado River. *Journal of Law and Economics* 23 (April): 111–34.

Clark, Robert E., Robert E. Beck, and Edward W. Clyde. 1972. *Waters and Water Rights*, vol. 5. Indianapolis, IN: Allen Smith Company.

Clyde, Steven E. 1989. Adapting to the Changing Demand for Water Use Through Continued Refinement of the Prior Appropriation Doctrine: An Alternative Approach to Wholesale Reallocation. *Natural Resources Journal* 29 (Spring): 435–55.

Colby, Bonnie G. 1988. Economic Impacts of Water Law: State Law and Water Market Development in the Southwest. *Natural Resources Journal* 28 (Fall): 721–49.

——. 1990. Transactions Costs and Efficiency in Western Water Allocation. *American Journal of Agricultural Economics* (December): 1184–92.

Colby, Bonnie, Mark A. McGinnis, and Ken Rait. 1989. "Procedural Aspects of State Water Law: Transferring Water Rights in the Western States." *Arizona Law Review* 31: 697–720.

Donohew, Zachary. 2009 Property Rights and Western United States Water Market. *Australia Journal of Agricultural and Resource Economics* 53 (1): 85–103.

Getches, David. 1984. *Water Law in a Nutshell*. St. Paul, MN: West Publishing Co.

——. "Water Resources: A Wider World." In *Natural Resources Policy and Law: Trends and Directions*, ed. Lawrence J. MacDonnell and Sarah F. Bates. Washington, DC: Island Press, 124–47.

Glennon, Robert J. 2009. *Unquenchable: America's Water Crisis and What To Do About It*. Washington, DC: Island Press.

Gould, George A. 1988. "Water Rights Transfers and Third-Party Effects." *Land and Water Law Review* 23: 1–41.

——. 1989. Transfer of Water Rights. *Natural Resources Journal* 29 (Spring): 457–77.

Gray, Brian E., Bruce C. Driver, and Richard W. Wahl. 1991. Economic Incentives for Environmental Protection. *Environmental Law* 21: 911–61.

Hanak, Ellen. 2003. *Who Should Be Allowed to Sell Water in California? Third-Party Issues and the Water Market*. San Francisco: Public Policy Institute of California.

Hartman, L. M., and Don Seastone. 1970. *Water Transfers: Economic Efficiency and Alternative Institutions*. Baltimore, MD: Johns Hopkins University Press.

Hayward, Steven. 1991. Muddy Waters. *Reason* (July): 46–47.

——. 1992. Wrong Place: Water in California. *Economist* (February): 24.

Huffman, James L. 1991. Letter to the Editor. *Environmental Law* 21: 2253.

——. 1994. Markets, Regulation, and Environmental Protection. *Montana Law Review* 55 (Summer): 425–35.

——. 2008. Speaking of Inconvenient Truths: A History of the Public Trust Doctrine. *Duke Environmental Law and Policy Forum* 18 (1).

Ingram, Helen, and Cy R. Oggins. 1992. The Public Trust Doctrine and Community Values in Water. *Natural Resources Journal* 32 (Summer): 515–37.

Jaunich, R. Prescott. 1994. The Environment, The Free Market, and Property Rights: Post-Lucas Privatization of the Public Trust. *Public Land Law Review* 15: 167–97.

Johnson, Ronald N., Micha Gisser, and Michael Werner. 1981. The Definition of a Surface Water Right and Transferability. *Journal of Law and Economics* 14 (October): 273–88.

Kinney, Clesson. 1912. *Law of Irrigation and Water Rights and Arid Region Doctrine of Appropriation of Waters*, vol. 1. San Francisco: Bender-Moss.

Libecap, Gary D. 2005. Rescuing Water Markets: Lessons from Owens Valley. *PERC Policy Series* no. 33. Jane S. Shaw, series editor. Bozeman, MT: PERC.

McCurdy, Charles W. 1976. Stephen J. Field and Public Land Law Development in California, 1850–1866: A Case Study of Judicial Resource Allocation in Nineteenth-Century America. *Law and Society Review* 10 (Winter): 235–66.

Meyers, Charles J., and Richard A. Posner. 1971. *Market Transfers of Water Rights*. Legal Study 4. Washington, DC: National Water Commission.

National Research Council. 1992. *Water Transfers in the West: Efficiency, Equity and the Environment*. Washington, DC: National Academy Press.

Public Policy Institute of California. 2003. Managing California's Water Market: Issues and Prospects. *Research Brief Issue* 74. www.ppic.org/content/pubs/rb/RB_703EHRB.pdf (accessed: May 10, 2011).

Reisner, Mark, and Sarah Bates. 1993. *Overtapped Oasis: Reform or Revolution for Western Water*. Washington, DC: Island Press.

Shupe, Steven J., Gary D. Weatherford, and Elizabeth Checchio. 1989. Western Water Rights: The Era of Reallocation. *Natural Resources Journal* 29 (Spring): 413–34.

Sprankling, John G. 1994. An Environmental Critique of Adverse Possession. *Cornell Law Review* 79 (May): 816–84.

State Water Resources Control Board. 1999. A Guide to Water Transfers. http://www.waterrights.ca.gov/watertransferguide.pdf (accessed: May 10, 2011).

Thompson, Barton. 1993. Institutional Perspectives on Water Policy and Markets. *California Law Review* 81: 671–764.

Tregarthen, Timothy D. 1977. The Market for Property Rights in Water. In *Water Needs*, ed. Ved P. Nanda. Boulder, CO: Westview Press, 139–51.

Trelease, Frank J. 1955. Preferences to the Use of Utility. *Rocky Mountain Law Review* 27: 133–60.

——. 1976. Alternatives to Appropriation Law. In Water Needs for the Future: Legal, Political, Economic, and Technological Issues in National and International Perspectives. *Denver Journal of International Law and Policy* 6 (Special issue): 283–305.

Wahl, Richard W. 1989. *Markets for Federal Water: Subsidies, Property Rights, and the Bureau of Reclamation*. Washington, DC: Resources for the Future.

Western Governors' Association Water Efficiency Working Group. 1987. *Water Efficiency: Opportunities for Action: Report to the Western Governors*. July. Denver, CO: Western Governors' Association.

Wilkinson, Charles F. 1991. In Memorium: Prior Appropriation 1848–1991. *Environmental Law* 21 (3): 5–28.

Willey, Zach. 1992. Behind Schedule and Over Budget: The Case of Markets, Water, and Environment. *Harvard Journal of Law and Public Policy* 15: 391–425.

6   The New Frontier

Abrams, Robert H. 1990. Water Allocation by Comprehensive Permit Systems in the East: Considering a Move Away from Orthodoxy. *Virginia Environmental Law Journal* 9: 257–258, 262, 263.

Adler, Robert W. 2008. Revisiting the Colorado River Compact: Time for a Change? *Utah Environmental Law Review* 28 (1): 19–47.

American Society of Civil Engineers. 2004. *Regulated Riparian Model Water Code*. Reston, VA: American Society of Civil Engineers.

Anderson, Terry L., and Ronald N. Johnson. 1983. The Problem of Instream Flows. *Economic Inquiry* 24 (October): 535–54.

Anderson, Terry L. and Donald R. Leal. 2001. *Free Market Environmentalism*. New York: Palgrave.

Anderson, Terry L., and Dean Lueck. 1992. Land Tenure and Agricultural Productivity on Indian Reservations. *Journal of Law and Economics* 35 (2): 427–54.

Associated Press. 2010. Chattahoochee River Brings Life, Conflict to Alabama, Georgia, Florida. *Everything Alabama*, June 20.

Ausness, Richard C. 1986. Water Rights, the Public Trust Doctrine, and the Protection of Instream Uses. *University of Illinois Law Review* (2): 407–37.

Bretsen, Stephen N., and Peter J. Hill. 2009. Water Markets as a Tragedy of the Anticommons. *William and Mary Environmental Law and Policy Review* 33 (3): 723–83.

Brooks, Robert, and Edwyna Harris. 2008. Efficiency Gains from Water Markets: Empirical Analysis of Watermove in Australia. *Agricultural Water Management* 95 (4): 391–99.

Butler, Lynda L. 1985. Allocating Consumptive Water Rights in a Riparian Jurisdiction: Defining the Relationship between Public and Private Interests. *University of Pittsburgh Law Review* 47 (1): 95–181.

Caponera, Dante A. 2007. *Principles of Water Law and Administration*. London: Taylor and Francis.

Choe, Olivia S. 2004. Appurtenancy Reconceptualized: Managing Water in an Era of Scarcity. *Yale Law Journal* 113 (8): 1909–53.

Christman, James N. 1998. Riparian Doctrine. In *Water Rights of the Eastern United States*, ed. Kenneth R. Wright. Denver, CO: American Water Works Association.

Coase, Ronald. 1960. The Problem of Social Cost. *Journal of Law and Economics* 3 (October): 1–44.

Dellapenna, Joseph W. 1991. Regulated Riparianism. In *Waters and Water Rights*, vol. 1, ed. Robert E. Beck, repl. vol. 2001. Charlottesville, VA: Lexus Law Publishing, chs. 6 and 7.

———. 2002. The Law of Water Allocation in the Southeastern States at the Opening of the Twenty-first Century. *University of Arkansas at Little Rock Law Review* 25: 9–88.

———. 2004. Special Challenges to Water Markets in Riparian States. *Georgia State University Law Review* 21 (2): 305–38.

Eggertsson, Thrainn. 2003. Open Access versus Common Property. In *Property Rights: Cooperation, Conflict, and Law*, ed. Terry L. Anderson and Fred S. McChesney. Princeton, NJ: Princeton University Press, 73–89.

Flood, Patricia K., and Kenneth R. Wright. 1998. Summary of Water Rights Law in the 31 Eastern States. In *Water Rights of the Eastern United States*. Denver, CO: American Water Works Association, 107–39.

Glennon, Robert J. 2009. *Unquenchable: America's Water Crisis and What To Do About It*. Washington, DC: Island Press.

Heller, Michael. 1998. The Tragedy of the Anticommons: Property in the Transition from Marx to Markets. *Harvard Law Review* 111: 621–88.

Huffman, James. 1997. Institutional Constraints in Transboundary Water-Marketing. In *Water Marketing: The Next Generation*, ed. Terry L. Anderson and Peter J. Hill. Lanham, MD: Rowman and Littlefield, 31–42.

Klein, Christine A., Mary Jane Angelo, and Richard Hamann. 2009. Modernizing Water Law: The Example of Florida. *Florida Law Review* 61 (3): 403–74.

Lueck, Dean. 1995. The Rule of First Possession and the Design of the Law. *Journal of Law and Economics* 38 (2): 393–436.

North Carolina Division of Water Resources. 2008. North Carolina Drought Management Advisory Council Activities Report 2008. www.ncdrought.org/documents/2008_annual_report.pdf (accessed: March 22, 2010).

Rose, Carol M. 1990. Energy and Efficiency in the Realignment of Common-Law Water Rights. *Journal of Legal Studies* 19 (2): 261–96.

Ruhl, J. B. 2005. Water Wars, Eastern Style: Divvying Up the Apalachicola-Chattahoochee-Flint River Basin. *Journal of Contemporary Water Research and Education* 131: 47–54.

Scarborough, Brandon. 2010. Environmental Water Markets: Restoring Streams Through Trade. *PERC Policy Series* no. 46. Bozeman, MT: PERC.

Sherk, George W. 1990. Eastern Water Law: Trends in State Legislation. *Virginia Environmental Law Journal* 9 (2): 287–321.

Smith, Henry E. 2000. Semicommon Property Rights and Scattering in the Open Fields. *Journal of Legal Studies* 29 (1): 131–69.

——. 2008. Governing Water: The Semicommons of Fluid Property Rights. *Arizona Law Review* 51 (2): 445–78.

Tarlock, A. Dan. 2004. Water Law Reform in West Virginia: The Broader Context. *West Virginia Law Review* 106: 495–538.

Wang, Hui, Rong Fu, Arun Kumar, and Wenhong Li. 2010. Intensification of Summer Rainfall Variability in the Southeastern United States during Recent Decades. *Journal of Hydrometeorology* 11: 1007–18.

Waye, Vicki, and Christina Son. 2010. Regulating the Australian Water Market. *Journal of Environmental Law* 22 (3): 431–59.

7   Buy That Fish a Drink

Adelsman, Hedia. 2003. *Washington Water Acquisition Program: Finding Water to Restore Streams.* March, 03-11-005. Olympia, WA: Washington Dept. of Ecology, Publications Distribution Center. online: www.ecy.wa.gov/pubs/0311005.

Amman, Dave. 2008. Telephone conversation with Dave Amman, hydrologist, Montana Department of Natural Resources and Conservation. December 1.

Anderson, Terry L., and Ronald N. Johnson. 1983. The Problem of Instream Flows. *Economic Inquiry* 24 (October): 535–54.

Anderson, Terry L., and Donald R. Leal. 1991 *Free Market Environmentalism.* Boulder, CO: Westview Press. 2nd ed. 2001, New York: Palgrave.

Arizona Department of Water Resources. 2009. *The Arizona Water Atlas*, vols. 2–8. www.adwr.state.az.us/dwr/Content/Find_by_Program/Rural_Programs/content/water_atlas/default.htm (accessed: June 29, 2010).

Arizona Water Protection Fund Commission. 2008. *Arizona Water Protection Fund: Protecting Arizona's River and Riparian Resources, Annual Report 2007–2008.* www.awpf.state.az.us/pubs/FY2008/FY-2008-Annual-Report-1.pdf (accessed: May 16, 2009).

Baden, John and Fort, Rodney. 1980. Natural Resources and Bureaucratic Predators. *Policy Review* 11 (Winter): 69–82.

Beeman, Josephine P. 1993. Instream Flow in Idaho. In *Instream Flow Protection in the West*, ed. Lawrence J. MacDonnell and Teresa A. Rice, rev. ed. Boulder, CO: Natural Resources Law Center, University of Colorado School of Law, ch. 13.

Benson, Reed D. 2006. Adequate Progress or Rivers Left Behind? Developments in Colorado and Wyoming Instream Flow Laws Since 2000. *Environmental Law* 36: 1283–310.

Boyd, Jesse A. 2003. Hip Deep: A Survey of State Instream Flow Law from the Rocky Mountains to the Pacific Ocean. *Natural Resources Journal* 43: 1151–1216.

California Bay–Delta Authority 2008. Environmental Water Account Program Plan Year 9. www.calwater.ca.gov/content/Documents/library/ProgramPlans/2008/Draft_EWA_Program_Plan_Year_9.pdf (accessed: July 22, 2010).

Case, Cale. 2003. Instream Flow Law has Run its Course. *Frontline* Summer 2003: 17.

Coase, Ronald H. 1974. The Lighthouse in Economics. *Journal of Law and Economics* 17 (October): 257–76.

Colorado Trout Unlimited. 2008. Colorado Water Project. http://www.cotrout.org/Conservation/ColoradoWaterProject/tabid/89/Default.aspx (accessed: May 5, 2009).

Colorado Water Conservation Board. 2008. Water Acquisitions. http://cwcb.state.co.us/StreamAndLake/WaterAcquisitions/ (accessed: August 21, 2008).

Columbia Basin Water Transactions Program. 2006. Stories from the Field. http://www.cbwtp.org/jsp/cbwtp/stories/stories.jsp?year=2006 (accessed: November 22, 2008).

Deschutes River Conservancy. 2008. History of the Deschutes River Conservancy. http://www.deschutesriver.org/About_Us/History/default.aspx (accessed: August 16, 2008).

Gray, Brian E. 1993. A Reconsideration of Instream Appropriative Water Rights in California. In *Instream Flow Protection in the West*, ed. Lawrence J. MacDonnell and Teresa A. Rice. Boulder, CO: Natural Resources Law Center, University of Colorado School of Law, 11–23.

Hawkes, Timothy. 2006. *When Rivers Run Dry: The Need for In-Stream Flow Reform in Utah*. Brochure (Summer). Bountiful, Utah: Utah Water Project and Trout Unlimited.

Huffman, James. 1983. Instream Water Use: Public and Private Alternatives. In *Water Rights: Scarce Resource Allocation, Bureaucracy, and the Environment*, ed. Terry L. Anderson. Cambridge, MA: Ballinger Press, 249–282.

Idaho Department of Water Resources. 2010a. Idaho Minimum Stream Flow Program. http://www.idwr.idaho.gov/waterboard/WaterPlanning/Minimum%20Stream%20Flow/PDFs/MSF_Brochure.pdf (accessed: August 13, 2010).

——. 2010b. State Protected Rivers. http://www.idwr.idaho.gov/waterboard/WaterPlanning/Protected%20Rivers/protected_rivers.htm (accessed: January 23, 2010).

McKinney, Matthew. 1993. Instream Flow Policy in Montana: A History and Blueprint for the Future. In *Instream Flow Protection in the West*. Natural Resources Law Center, University of Colorado School of Law, revised edition 1993: 15-1–15-58.

Neuman, Janet C., and Cheyenne Chapman. 1999. Wading Into the Water Market: The First Five Years of the Oregon Water Trust. *Journal of Environmental Law and Litigation* 14: 135–84.

Oregon Water Resources Department. 2008. Oregon Water Resources Department Summary Presentation: Statewide Roundtables Fall 2008. http://water.oregonstate.edu/roundtables/download/OWRD_Presentation.pdf (accessed: December 2, 2008).

Pilz, Robert D. 2006. At the Confluence: Oregon's Instream Water Rights Law in Theory and Practice. *Environmental Law* 36: 1383–420.

Slattery, Kenneth O., and Robert F. Barwin. 1993. Protecting Instream Resources in Washington State. In *Instream Flow Protection in the West*, ed. Lawrence J. MacDonnell and Teresa A. Rice. Boulder, CO: Natural Resources Law Center, University of Colorado School of Law, 371–89.

Sterne, Jack. 1997. Instream Rights and Invisible Hands: Prospects for Private Instream Water Rights in the Northwest. *Environmental Law* 27 (1): 203–43.

The Nature Conservancy (TNC). 2006. *Annual Report*. Phoenix, AZ: The Nature Conservancy in Arizona. www.nature.org/wherewework/northamerica/states/arizona/files/ar_2006_web.pdf (accessed: May 16, 2011).

Trout Unlimited. 2008. Idaho: Restoring Stream Flows in Key River Basins. http://www.tu.org/conservation/western-water-project/idaho/id-restoring-stream-flows-in-key-river-basins (accessed: September 4, 2009).

Washington Department of Ecology. 2007. Mitigation Guide for Future Outdoor Water Use in the Walla Walla Basin. www.ecy.wa.gov/pubs/0711032.pdf (accessed: July 2, 2008).

——. 2008. Map of Existing and Future Instream Flow Rules and USGS Gages, January 2008. www.ecy.wa.gov/programs/wr/instream-flows/Images/irpp_wrp/wsisf0108-usgs.pdf (accessed: 14 September 2008).

Washington Water Trust and Washington Department of Ecology. 2007. Mitigation Guide for Future Outdoor Water Use in the Walla Walla Basin. Publication no. 07-11-032. http://www.ecy.wa.gov/pubs/0711032.pdf (accessed: September 14, 2009).

8 Good to the Last Drop

Abdalla, Charles, Tatiana Borisova, Doug Parker, and Kristen Saacke Blunk. 2007. Water Quality Credit Trading and Agriculture: Recognizing the Challenges and Policy Issues Ahead. *Choices* 22 (2): 117–24.

Ackerman, Bruce A., and Richard B. Stewart. 1988. Reforming Environmental Law: The Democratic Case for Market Incentives. *Columbia Journal of Environmental Law* 13: 171–99.

Adler, Jonathan H. 2002. Fables of the Cuyahoga: Reconstructing a History of Environmental Protection. *Fordham Environmental Law Journal* 14: 89–146.

Bartfeld, Esther. 1993. Point-Nonpoint Source Trading: Looking Beyond Potential Cost Savings. *Environmental Law* 23: 43–106.

Bate, Roger. 1994. Water Pollution Prevention: A Nuisance Approach. *Economic Affairs* 14 (3): 13–14.

Battle, Jackson B., and Maxine I. Lipeles. 1993. *Environmental Law*, vol. 2, 2nd ed. Cincinnati, OH: Anderson Publishing Co.

Beck, Robert E., and P. Goplerud. 1988. *Water Pollution and Water Quality: Legal Controls*. Charlottesville, VA: The Michie Company.

Black, Henry Campbell. 1983. *Black's Law Dictionary*. St. Paul, MN: West Publishing Co.

Bonine, John E., and Thomas O. McGarity. 1984. *The Law of Environmental Protection*. St. Paul, MN: West Publishing Co.

Breetz, Hanna, Karen Fisher-Vanden, Laura Garzon, Hannah Jacobs, Kailin Kroetz, and Rebecca Terry. 2004. *Water Quality Trading and Offset Initiatives in the US: A Comprehensive Survey*. Dartmouth College. http://www.dep.state.fl.us/water/watersheds/docs/ptpac/DartmouthCompTradingSurvey.pdf (accessed: September 23, 2009).

Brown, Gardner M., and Ralph W. Johnson. 1984. Pollution Control by Effluent Charges: It Works in the Federal Republic of Germany, Why Not in the U.S? *Natural Resources Journal* 24 (October): 929–66.

Brubaker, Elizabeth. 1995. *Property Rights in the Defense of Nature*. London: Earthscan Publications.

Buchele, Thomas C. 1986. State Common Law Actions and Federal Pollution Control Statutes: Can They Work Together? *University of Illinois Law Review*: 609–44.

Chesapeake Bay Foundation. 2009. The Chesapeake Clean Water and Ecosystem Restoration Act of 2009. HR 3852/S 1816. www.cbf.org/Document.Doc?id=398 (accessed: June 5, 2010).

Clean Water Services. 2008. Annual Report 2007–2008. www.cleanwaterservices.org/Content/Documents/About%20Us/Annual%20Report%202007-2008.pdf (accessed: May 13, 2010).

Connecticut Department of Environmental Protection (CTDEP). 2003. Managing Environmental Compliance in Connecticut, November 2003. www.ct.gov/dep/lib/dep/enforcement/newsletter/nov03.pdf (accessed: May 2, 2010).

———. 2009. Report of the Nitrogen Credit Advisory Board for Calendar Year 2008: To the Joint Standing Environment Committee of the General Assembly. ct.gov/dep/lib/dep/water/lis_water_quality/nitrogen_control_program/nitrogen_report_2008.pdf (accessed: February 3, 2010).

Conservation Foundation. 1987. *State of the Environment: A View Toward the Nineties.* Washington, DC: The Conservation Foundation.

Copeland, Claudia. 2003. EPA's Water Quality Trading Policy. *Congressional Research Service Report* RS21403. stuff.mit.edu/afs/sipb/contrib/wikileaks-crs/wikileaks-crs-reports/RS21403.pdf (accessed: December 3, 2009).

Dasgupta, Susmita, Benoit Laplante, Hua Wang, and David Wheeler. 2002. Confronting the Environmental Kuznets curve. *Journal of Economic Perspectives* 16: 147–68.

Davidson, John H. 1989. Commentary: Using Special Water Districts to Control Nonpoint Sources of Water Pollution. *Chicago-Kent Law Review* 65: 503–18.

Davis, Peter N. 1993. Law and Fact Patterns in Common Law Water Pollution Cases. *Missouri Environmental Law and Policy Review* 1 (Summer): 3–16.

Ember, Lois R. 1992. Clean Water Act is Sailing a Choppy Course to Renewal. *Chemical and Engineering News* 70 (February 17): 18–22.

Fish Legal. 2010. Using the Law to Protect Fisheries and Angling. www.fishlegal.net/page.asp?section=166&sectionTitle=What+does+Fish+Legal+do%3F (accessed: May 22, 2010).

Food and Agriculture Organization. 2000. *Fertilizer Requirements in 2015 and 2030.* Rome: FAO.

Foran, Jeffery A., Peter Butler, Lisa B. Cleckner, and Jonathan W. Bulkley. 1991. Regulating Nonpoint Source Pollution in Surface Waters: A Proposal. *Water Resources Bulletin* 27 (3): 479–83.

Gardner, B. Delworth. 1995. *Plowing Ground in Washington.* San Francisco: Pacific Research Institute for Public Policy Research.

Hall, John, and Ciannat Howett. 1994. Albemarle-Pamlico: Case Study in Pollutant Trading. *EPA Journal* (Summer): 27–29.

Hanson, David J. 1994. Reauthorization of Clean Water Act Remains Debatable, Raises Objections. *National Journal* (March 19): 37–39.

Houck, Oliver A. 2002. *The Clean Water Act TMDL Program: Law, Policy, and Implementation,* 2nd ed. Washington, DC: Environmental Law Institute.

Kieser and Associates. 2004. Preliminary Economic Analysis of Water Quality Trading Opportunities in the Great Miami River Watershed, Ohio. Prepared for: The Miami Conservancy District. www.miamiconservancy.org/water/documents/Executive Summary.pdf (accessed: June 23, 2010).

King, Dennis M. 2005. Crunch Time for Water Quality Trading. *Choices* 20 (1): 71–76.

King, Dennis M., and Peter J. Kuch. 2003. Will Nutrient Trading Ever Work? An Assessment of Supply and Demand Problems and Institutional Obstacles. *Environmental Law Reporter* 33 (5): 10352–68.

Letson, David. 1992. Point/Nonpoint Source Pollution Reduction Trading: An Interpretive Survey. *Natural Resources Journal* 32 (Spring): 219–32.

Lewin, Jeff L. 1990. Boomer and the American Law of Nuisance: Past, Present, and Future. *Albany Law Review* 54: 189–300.

Malik, Arun S., David Letson, and Stephen R. Crutchfield. 1993. Point/Nonpoint Source Trading of Pollution Abatement: Choosing the Right Trading Ratio. *American Journal of Agricultural Economics* 75 (November): 959–67.

Meiners, Roger E., and Bruce Yandle. 1994a. Clean Water Legislation: Reauthorize or Repeal? In *Taking the Environment Seriously*, ed. Roger Meiners and Bruce Yandle. Lanham, MD: Rowman and Littlefield, ch. 4.

———. 1994b. *Reforming the Clean Water Act*. Washington, DC: Manufacturers' Alliance for Productivity and Innovation.

Moon, David. 1995. Personal telephone conversation between David Moon, attorney for the Stevensons, and author, May 17.

Morgan, Cynthia, and Ann Wolverton. 2005. Water Quality Trading in the United States. *National Center for Environmental Economics Working Paper Series* 05–07, June.

Niemi, Ernie, Kristin Lee, and Tatiana Raterman. 2007. Net Economic Benefits of Using Ecosystem Restoration to Meet Stream Temperature Objectives. *Proceedings of the Water Environment Federation* 9: 611–19.

Office of Management and Budget. 2010. Budget of the U.S. Government Fiscal Year 2011. www.whitehouse.gov/omb/budget/fy2011/assets/budget.pdf (accessed: January 22, 2011).

Paul, Michael J., and Judy L. Meyer. 2001. Streams in the Urban Landscape. *Annual Review of Ecology and Systematics* 32: 333–65.

Powers, Ann. 2003. The Current Controversy Regarding TMLDS: Pollutant Trading. *Vermont Journal of Environmental Law* 4 (1): 9–41.

Prosser, W. L. and Keeton, R. E. 1984. *Prosser and Keeton on the Law of Torts*. 5th edition. St. Paul, MN: West Publishing Co.

Riggs, David W. 1993. *Market Incentives for Water Quality: A Case Study of the Tar-Pamlico River Basin, North Carolina*. Clemson, SC: Center for Policy Studies.

Rodgers, William H. Jr. 1994. *Environmental Law*. St. Paul, MN: West Publishing Co.

RNC Consulting. 2003. Bear Creek Watershed Report 2002: Annual Report and Water Quality Summary Sheets. A Report Prepared for the Bear Creek Watershed Association. www.bearcreekwatershed.org/Monitoring%20Program/Annual%20Reports/2005%20Bear%20Creek%20Annual%20Report.pdf (accessed: March 22, 2009).

Rubin, Debra K., Mary B. Powers, Housley Carr, and David B. Rosenbaum. 1993. A Whole Lot of Planning Going On. *ENR* 231 (September): 38–44.

Selman, Mindy, and Suzie Greenhalgh. 2009. Eutrophication: Sources and Drivers of Nutrient Pollution. *WRI Policy Note, Water Quality: Eutrophication and Hypoxia*, no. 2, June.

Selman, Mindy, Suzie Greenhalgh, Evan Branosky, Cy Jones, and Jenny Guiling. 2009. Water Quality Trading Programs: An International Overview. *WRI Issue Brief*, no. 1, March.

Toft, Dennis M. 1994. EPA and Affected States to Step up Enforcement of the Clean Water Act. *National Law Journal* (June 20): C6–11.

United Nations Development Programme. 2009. *Human Development Report 2006: Beyond Scarcity: Power, Poverty and the Global Water Crisis*. New York: Palgrave Macmillan.

U.S. Congressional Research Service. 2008. Water Quality Issues in the 110th Congress: Oversight and implementation. www.nationalaglawcenter.org/assets/crs/RL33800.pdf (accessed: May 22, 2010).

———. 2008a. Wetlands: An Overview of Issues, 2010. www.ncseonline.org/NLE/CRSreports/08Aug/RL33483.pdf (accessed: July 2).

U.S. Environmental Protection Agency (EPA). 1996. Draft Framework for Watershed-Based Trading. EPA 800-R-96-001 (May 1996).

——. 2003. Proposed Water Quality Trading Policy. http://water.epa.gov/type/watersheds/ proptradepolicy.cfm (accessed: September 13, 2009).

——. 2007. Water Quality Trading Toolkit for Permit Writers. Office of Wastewater Management, Water Permit Division. EPA 833-R-07-004.

——. 2008. Clean Watersheds Needs Survey 2004: Report to Congress. www.epa.gov/ owm/mtb/cwns/2004rtc/toc.htm (accessed: August 2, 2010).

——. 2008a. Water Quality Trading Evaluation Final Report. www.epa.gov/evaluate/ pdf/wqt.pdf (accessed: August 2, 2010).

——. 2009. Restoring the Long Island Sound While Saving Money: Lessons in Innovation and Collaboration. www.epa.gov/owow/TMDL/tmdlsatwork/pdf/long_island_sound_ byte.pdf (accessed: February 3, 2010).

——. 2010. State and Individual Trading Programs. water.epa.gov/type/watersheds/ trading/tradingmap.cfm (accessed: March 15, 2010).

——. 2010a. Clean Water Act Section 319(h) Grant Funds History. www.epa.gov/ owow/nps/319hhistory.html (accessed: October 18, 2010).

U.S. Water News. 1994. Permits Can Shield Pollution Liability. U.S. Water News 11 (May): 1 and 2.

——. 1994a. Farmers Sowing More "Green Stripes." U.S. Water News 11 (July): 4.

——. 1995. Litigation Threat to Force Clean Water Act Renewal. U.S. Water News 11 (April): 1 and 7.

Willey, Zach. 1990. Environmental Quality and Agricultural Irrigation: Reconciliation Through Economic Incentives. In Agricultural Salinity Assessment and Management, ed. Kenneth K. Tanjii. New York: American Society of Civil Engineers, 391–425.

Woodward, Richard T. 2006. The Challenge of Transaction Costs. Presentation at a Workshop on Environmental Credits Generated Through Land-Use Changes: Challenges and Approaches, March 8–9, 2006. Baltimore, MD. http://www.envtn.org/ LBcreditsworkshop/Transaction_Costs_Intro.pdf (accessed: March 30, 2010).

World Resources Institute. 1990. World Resources 1990–91. New York: Oxford University Press.

——. 1993. The 1993 Information Please Environmental Almanac. New York: Houghton Mifflin Company.

Yandle, Bruce. 1994. Community Markets to Control Agricultural Nonpoint Source Pollution. In Taking the Environment Seriously, ed. Roger E. Meiners and Bruce Yandle. Lanham, MD: Rowman and Littlefield, 185–207.

9   The Race to Pump

Anderson, Terry L., and P. J. Hill. 1981. Establishing Property Rights in Energy: Efficient v. Inefficient Processes. Cato Journal 1 (1): 87–105.

Anderson, Terry L., Oscar R. Burt, and David T. Fractor. 1983. Privatizing Groundwater Basins: A Model and Its Application. In Water Rights: Scarce Resource Allocation, Bureaucracy, and the Environment. San Francisco: Pacific Institute for Public PolicyResearch, 223–48.

Balleau, W. Peter. 1988. Water Appropriation and Transfer in a General Hydrogeologic System. Natural Resources Journal 28 (2): 269–91.

Barberis, Julio. 1991. The Development of International Law of Transboundary Groundwater. Natural Resources Journal 31 (Winter): 167–86.

California Department of Water Resources. 2006. Cummings Valley Groundwater Basin. California's Groundwater Bulletin 118. www.water.ca.gov/pubs/groundwater/ bulletin_118/basindescriptions/5-27.pdf (accessed: March 3, 2010).

Dennehy, Kevin F. 2000. High Plains Regional Ground-water Study: U.S. Geological Survey Fact Sheet FS-091-00. co.water.usgs.gov/nawqa/hpgw/factsheets/DENNEHYFS1.html (accessed: August 17, 2009).

Delleur, Jacques W. 2007. *The Handbook of Groundwater Engineering*. Boca Raton, FL: CRC Press.

Escalante Valley Water Users Association. 2011. Escalante Valley Water Users Ground Water Management Plan. evwua.org/pdf/PP2011.pdf (accessed: April 20, 2011).

Garner, Eric L., Michelle Ouellette, and Richard L. Sharff Jr. 1994. Institutional Reforms in California Groundwater Law. *Pacific Law Journal* 25: 1021–52.

Gates, John M. 1969. Repayment and Pricing in Water Policy: A Regional Economic Analysis with Particular Reference to the Tehachapi–Cummings County Water District. Ph.D. dissertation, Department of Agricultural Economics, University of California–Berkeley.

Glennon, Robert. 2002. *Water Follies: Groundwater Pumping and the Fate of America's Fresh Waters*. Washington, DC: Island Press.

Glennon, Robert, and Thomas Maddock III. 1994. In Search of Subflow: Arizona's Futile Effort to Separate Groundwater from Surface Water. *Arizona Law Review* 36: 567–610.

Hansen, J. 2010. It Takes a District: Utah Landowners Control Groundwater Use, Escalante Valley Citizens Plan to Save Their Declining Aquifer. *High Country News* May 10, 2010.

Heath, Ralph C. 1988. Groundwater. In *Perspectives on Water: Uses and Abuses*, ed. David H. Speidel, Lon C. Ruedisili, and Allen F. Agnew. New York: Oxford University Press, 73–89.

Kenny, Joan F., Nancy L. Barber, Susan S. Hutson, Kristin S. Linsey, John K. Lovelace, and Molly A. Maupin. 2009. Estimated Use of Water in the United States in 2005. *U.S. Geological Survey Circular* 1344.

Libecap, Gary D. 2008. Transaction Costs, Property Rights, and the Tools of the New Institutional Economics: Water Rights and Water Markets. In *The New Institutional Economics: A Guidebook*, ed. Eric Brousseau and Jean-Michel Glachant, New York: Cambridge University Press, 272–91.

Lipson, Albert J. 1978. *Efficient Water Use in California: The Evolution of Groundwater Management in Southern California*. 4-2387/2-CSA/RF, November. Washington, DC: The Rand Corporation.

Little, Jane B. 2009. Saving the Ogallala Aquifer. *Scientific American* 19 (March): 32–39.

Maupin, Molly A., and Nancy L. Barber. 2005. Estimated Withdrawals from Principal Aquifers in the United States, 2000. *U.S. Geological Survey Circular* 1279.

McGuire, V. L. 2007. Changes in Water Level and Storage in the High Plains Aquifer, Predevelopment to 2005. *U.S. Geological Survey Fact Sheet* 2007–3029.

Murray, Paula C., and Frank B. Cross. 1992. The Case for a Texas Compulsory Unitization Statute. *Saint Mary's Law Journal* 23: 1099–154.

Oklahoma Department of Environmental Quality. 2010. *Water Quality in Oklahoma: 2010 Integrated Report*. www.deq.state.ok.us/WQDnew/305b_303d/2010_draft_integrated_report.pdf (accessed: March 20, 2011).

Oklahoma Water Resources Board. 2011. Groundwater Studies. www.owrb.ok.gov/studies/groundwater/groundwater.php (accessed: April 2, 2011).

Provencher, Bill, and Oscar Burt. 1993. A Private Property Rights Regime for the Commons: The Case of Groundwater. *American Journal of Agricultural Economics* 76 (November): 875–88.

Reilly, Thomas E., Kevin F. Dennehy, William M. Alley, and William L. Cunningham. 2008. Ground-Water Availability in the United States. *U.S. Geological Survey Circular* 1323. pubs.usgs.gov/circ/1323/ (accessed: January 4, 2010).

Roberts, Rebecca S., and Sally L. Gros. 1987. The Politics of Groundwater Management Reform in Oklahoma. *Groundwater* 25 (September–October): 535–44.

Sax, Joseph L., and Robert H. Abrams. 1986. *Legal Control of Water Resources: Cases and Materials*. St. Paul, MN: West Publishing Company.

Smith, Vernon L. 1977. Water Deeds: A Proposed Solution to the Water Valuation Problem. *Arizona Review* 26 (January): 7–10.

South Florida Water Management District. 2009. Climate Change and Water Management in South Florida. Interdepartmental Climate Change Group. mytest.sfwmd.gov/ portal/page/portal/xrepository/sfwmd_repository_pdf/climate_change_and_water_manag ement_in_sflorida_12nov2009.pdf (accessed: March 2, 2011).

Stewart, B. 2003. Aquifer Ogallala. In *Encyclopedia of Water Science*. New York: Marcel Dekker, 43–44.

Stoebuck, William B., and Dale A. Whitman. 2000. *The Law of Property [1984]*. 3rd ed. St. Paul, MN.: West Brook.

Tehachapi–Cummings County Water District. 2008. *Thirty-Fifth Annual Watermaster Report, Tehachapi Basin*. Filed April 30, 2009 in the Superior Court of the State of California for the County of Kern.

Tehachapi Soil Conservation District. 1969. *Tehachapi Project Report*, March 31.

Trelease, Frank J. 1976. Developments on Groundwater Law. In *Advances in Groundwater "Mining" in the Southwestern States*, ed. Z. A. Saleem. Minneapolis, MN: American Water Resources Association. 271–78.

Weiser, Matt. 2009. Feds Document Shrinking San Joaquin Valley Aquifer. *Sacramento Bee*, July 13, 3A.

10   You Can't Keep a Good Market Down

ACIL Tasman, in association with Freehills. 2004. An Effective System of Defining Water Property Titles. *Research Report, Land and Water Australia*, Canberra, Australia.

Australian Bureau of Statistics. 2008. *Experimental Estimates of Industry Multifactor Productivity, 2007–08*. Catalogue no. 5260.0.55.0022.

Australian Department of Sustainability, Environment, Water, Population and Communities. 2010. Water for the Future: National Water Market System. www.environment.gov.au/ water/policy-programs/nwms/pubs/nwms-factsheet.pdf (accessed: March 22, 2011).

Barlow, Maude. 2007. *Blue Covenant*. New York: New Press.

Barringer, Felicity. 2010. Las Vegas's Worried Water Czar. *New York Times*, September 28.

Barrionuevo, Alexei. 2009. Chilean Town Withers in Free Market for Water. *New York Times*, March 15, A12.

Bauer, Carl J. 1997. Bringing Water Markets Down to Earth: The Political Economy of Water Rights in Chile, 1976–95. *World Development* 25 (5): 639–56.

———. 2004. Results of Chilean Water Markets: Empirical Research Since 1990. *Water Resources Research* 40.

———. 2005. In the Image of the Market: The Chilean Model of Water Resources Management. *International Journal of Water* 3 (2): 146–65.

Bechtel Corporation. 2005. Cochabamba and the Aguas del Tunari Consortium. www.bechtel.com/newsarticles/65.asp (accessed: 18 October 2010).

Bennett, Jeff. 2005. *The Evolution of Markets for Water: Theory and Practice in Australia*. Cheltenham, UK: Edward Elgar.

Bonnardeaux, David. 2009. The Cochabamba "Water War": An Anti-Privatisation Poster Child? Published collaboratively by IPN and 18 other think tanks around the world, on March 16, 2009. http://www.policynetwork.net/uploaded/pdf/Cochabamba_March09. pdf (accessed: June 18, 2010).

Brennan, D. and M. Scoccimarro. 2000. Issues in defining property rights to improve Australian water markets. *The Australian Journal of Agricultural and Resource Economics* 43 (1): 69–89.

Connor, Michael L. 2009. Statement of Michael L. Connor, Commissioner Bureau of Reclamation U.S. Department of the Interior Before the Energy and Natural Resources Committee Subcommittee on Water and Power. U.S. Senate on S. 1759 Water Transfer Facilitation Act of 2009, November 5. http://www.usbr.gov/newsroom/testimony/detail.cfm?RecordID=1521 (accessed: June 22, 2010).

Council of Australian Governments (COAG). 2004. Intergovernmental Agreement on a National Water Initiative. www.nwc.gov.au/resources/documents/Intergovernmental-Agreement-on-a-national-water-initiative2.pdf (accessed: June 24, 2010).

Davies, Peter E., John H. Harris, Terry J. Hillman, and Keith F. Walker. 2008. *Sustainable Rivers Audit Report 1: A Report on the Ecological Health of Rivers in the Murray–Darling basin 2004–2007.* Prepared by the Independent Sustainable Rivers Audit Group for the Murray–Darling Basin Ministerial Council. Canberra, Australia: Murray–Darling Basin Commission.

Donoso, Guillermo. 2006. Water Markets: Case Study of Chile's 1981 Water Code. *Ciencia e Investigacion Agraria* 33 (2): 157–71.

Dourojeanni, Axel, and Andrei Jouravlev. 1999. El Código De Aguas De Chile: Entre La Ideología Y La Realidad. *Recursos Naturales e Infrastructura* 3. Santiago, Chile: CEPAL.

Figueroa del Río, Luis Simon. 1995. *Asignación y Distribución de las Aguas Terrestres.* Santiago, Chile: Editorial Universidad Gabriela Mistral.

Garry, Thomas. 2007. Water Markets and Water Rights in the United States: Lessons from Australia. *Macquarie Journal of International and Comparative Environmental Law* 4: 23–60.

Gazmari, Renato, and Mark Rosengrant. 1996. Chilean Water Policy: The Role of Water Rights, Institutions, and Markets. *Water Resources Development* 12 (1): 33–48.

Grafton, R. Q., C. Landry, G. D. Libecap, and R. J. O'Brien. 2010. Water Markets: Australia's Murray–Darling Basin and the U.S. Southwest. National Bureau of Economic Research Working Paper 15797.

Haisman, Brian. 2005. Impacts of Water Rights Reform in Australia. In *Water Rights Reform: Lessons for Institutional Design*, ed. Bryan Randolph Bruns, Claudia Ringler, and Ruth Meinzen-Dick. Washington, DC: International Food Policy Research Institute, 113–52.

Hearne, Robert R., and Guillermo Donoso. 2005. Water Institutional Reforms in Chile. *Water Policy* 7: 53–69.

Israel, Debra K. 2007. Impact of Increased Access and Price on Household Water Use in Urban Bolivia. *The Journal of Environment and Development* 16 (1): 58–83.

Johnson, Robert. 2008. United States Department of the Interior letter to U.S. Congress. www.owrc.org/FedProgram/BOR_TitleTransfer_DraftBill.pdf (accessed: October 12, 2010).

Las Vegas Valley Water District. 2011. Operating and Capital Budget, Fiscal Year Ending June 30, 2011. www.lvvwd.com/about/financial_budget.html (accessed: March 13, 2011).

Murawski, John. 2011. Private Water Utility Wants Hefty Rate Hike. *Charlotte Observer*, February 15.

Murray–Darling Basin Authority (MDBA). 2010a. *Annual Report 2009–10.* Publication no. 110/10. Canberra, Australia: Murray–Darling Basin Authority.

———. 2010b. *Guide to the Proposed Basin Plan: Overview.* Canberra, Australia: Murray–Darling Basin Authority.

Murray–Darling Basin Commission (MDBC). 2008. Review of Cap Implementation 2006/07; Report of the Independent Audit Group. Pub no. 27-08 (accessed: December 3, 2010).

National Water Commission. 2010. *Australian Water Markets Report 2009–10*. Canberra, Australia: National Water Commission.

Nickson, Andrew, and Claudia Vargas. 2002. The Limitations of Water Regulation: The Failure of the Cochabamba Concession in Bolivia. *Bulletin of Latin American Research* 21 (1): 99–120.

North, Mike. 2010. Recall Election: Livingston's Mayor Daniel Varela and Councilwoman Martha Nateras Voted Out of Office. *Merced Sun-Star*, August 31. www.mercedsunstar.com/2010/08/31/1551581/livingstons-mayor-daniel-varela.html (accessed: December 13, 2010).

Quiggin, John. 2001. Environmental Economics and the Murray Darling River System. *Australian Journal of Agricultural and Resource Economics* 45 (1): 67–94.

Schleyer, Renato Gazmuri. 1994. Chile's Market-Oriented Water Policy: Institutional Aspects and Achievements. In *Water Policy and Water Markets: Selected Papers and Proceedings from the World Bank's Ninth Annual Irrigation and Drainage Seminar, Annapolis, Maryland, December 8–10, 1992*, ed. Guy Le Moigne, D. William Easter, Walter J. Ochs, and Sandra Giltner. Technical Paper, no. 249. Washington, DC: World Bank.

Schultz, Jim. 2003. Bolivia: The Water War Widens. *NACLA Report on the Americas* 26 (4): 34–47.

———. 2010. The Cochabamba Water Revolt, Ten Years Later. *Yes! Magazine*, April 20.

Southern Nevada Water Authority. 2010. Clark, Lincoln, and White Pine Counties Groundwater Development Project Conceptual Plan of Development. April. http://www.snwa.com/assets/pdf/gdp_concept_plan.pdf (accessed: February 13, 2011).

———. 2011. Major Construction and Capital Plan. February. http://www.snwa.com/assets/pdf/cip_mccp.pdf (accessed: March 3, 2011).

Sturgess, Gary L., and Michael Wright. 1993. *Water Rights in Rural New South Wales: The Evolution of a Property Rights System*. Sydney, Australia: The Centre for Independent Studies.

United Nations Environmental Programme (UNEP). 2007. *Global Environmental Outlook 4: Environment for Development*. Valletta, Malta: Progress Press Ltd.

U.S. Bureau of Reclamation. 2006. The Bureau of Reclamation's Title Transfer Program: Evaluation and Lessons Learned. Draft Report on Bureau of Reclamation's Title Transfer Efforts. www.usbr.gov/excellence/Drafts/43HOAI28.pdf (accessed: June 22, 2010).

U.S. Department of Homeland Security and Environmental Protection Agency (EPA). 2007. Critical Infrastructure and Key Resources Sector-Specific Plan as Input to the National Infrastructure Protection Plan. www.ct.gov/dph/lib/dph/drinking_water/pdf/water_sector-specific_plan.pdf (accessed: June 4, 2010).

U.S. Department of the Interior. 2009. Budget Justifications and Performance Information Fiscal Year 2009. Central Utah Project Completion Act. http://www.doi.gov/budget/2009/data/greenbook/FY2009_CUPCA_Greenbook.pdf (accessed: June 3, 2010).

U.S. Environmental Protection Agency (U.S. EPA). 2002. *The Clean Water and Drinking Water Infrastructure Gap Analysis*. EPA-816-R-02-020, September.

U.S. General Accounting Office (U.S. GAO). 2002. Water Infrastructure: Information on Financing, Capital Planning, and Privatization. GAO-02-764. Report to Congressional Requesters. August.

U.S. Water News. 2006. Las Vegas to Restrict Residents' Water Use. *U.S. Water News Online*, March.

Veolia Water. 2011. Finding the Blue Path for A Sustainable Economy. White Paper. March. www.veoliawaterna.com/north-america-water/ressources/documents/1/19979,IFPRI-White-Paper.pdf (accessed: August 30, 2011).

Vesilind, Pritt J. 2003. The Driest Place on Earth. *National Geographic*. August.

Walton, Brett. 2010. The Price of Water: A Comparison of Water Rates, Usage in 30 U.S. Cities. Circle of Blue. www.circleofblue.org/waternews/2010/world/the-price-of-water-a-comparison-of-water-rates-usage-in-30-u-s-cities/ (accessed: October 21, 2010).

Wilkinson, Charles. 1991. In Memoriam, Prior Appropriation 1848–1991. *Environmental Law* 3 (1): v–xviii.

Williams, Sandy, and Sarah Carriger. 2005. Water and Sustainable Development: Lessons from Chile. *Policy Brief* 2. Santiago, Chile: Global Water Partnership South America.

INDEX

Page references in *italics* denotes a table/figure.